Ingenious Women

Ingenious Women

From Tincture of Saffron to Flying Machines

DEBORAH JAFFÉ

Foreword by Sandi Toksvig

SUTTON PUBLISHING

First published in the United Kingdom in 2003 by
Sutton Publishing Limited · Phoenix Mill
Thrupp · Stroud · Gloucestershire · GL5 2BU

This paperback edition first published in 2004

British Library Cataloguing in Publication Data
A catalogue record for this book is available from the British Library.

ISBN 0-7509-3031-4

Typeset in 11/13.5 pt Garamond.
Typesetting and origination by
Sutton Publishing Limited.
Printed and bound in England by
J.H. Haynes & Co. Ltd, Sparkford.

To
These Ingenious Women

Contents

Foreword

The old saying is that 'Necessity is the mother of invention'. This may be true but it leaves out the fact that the inventor may also be somebody's mother. Years ago, when I was studying anthropology at university one of my female professors held up a photograph of an antler bone with twenty-eight markings on it. 'This,' she said, 'is alleged to be man's first attempt at a calendar.' We all looked at the bone in admiration. 'Tell me,' she continued, 'what man needs to know when twenty-eight days have passed? I suspect that this is woman's first attempt at a calendar.'

There are a lot of things which no one can prove. I firmly believe that women invented agriculture. If the men were all busy being he-man hunters who had the time to pop some seeds in the ground and watch them grow? The women left behind holding the baby. Studies of modern hunter-gatherer societies show that the hunters (the men) tend to bring in only 10 per cent of the group's food while the gatherers (the women) make up the remaining 90 per cent. Who gets all the publicity? The big butch hunter.

Part of the problem is that women and their achievements have been the silent story of history for too long. It won't surprise you to learn that a woman invented the chocolate chip cookie (Ruth Wakefield in 1930) or the dishwasher (Josephine Cochran in 1872) or even the disposable nappy (Marion Donovan in 1950) but did you know that it was women who made the first computer work? That it was a woman who found the first computer bug? (Grace Murray Hopper in 1947. Grace went on to become the first rear admiral

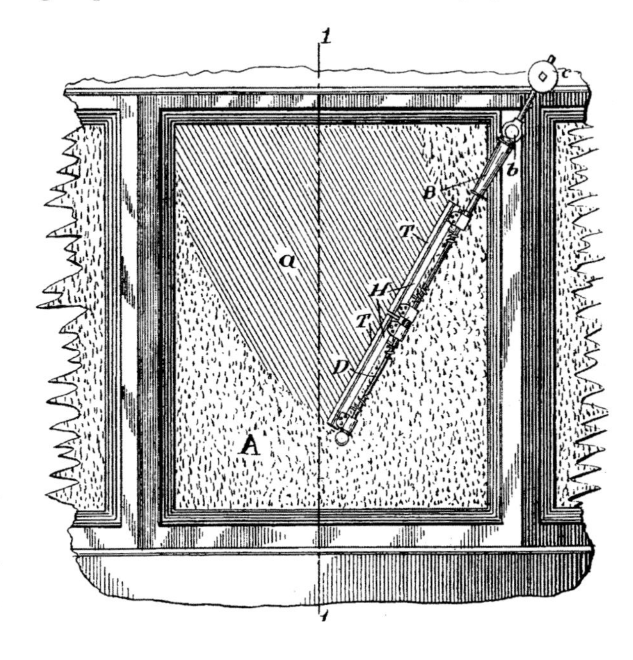

Part of Mary Anderson's windscreenwiper patent, 1903.

in the United States Navy.) How about Stephanie Kwolek who invented Kevlar, a steel-like fibre used in tyres, crash helmets and bullet-proof vests?

Some of the stories may never be told. Many people think that Sybilla Masters may have been the first American woman inventor. In 1712 she is said to have invented a new corn mill but she couldn't patent it because women weren't allowed to register. Three years later a patent was successfully filed in her husband's name.

This is a book which needed to be written. It is time to celebrate the way in which women have changed and improved the world. My only regret is that I didn't think to write it myself.

Sandi Toksvig

Preface

Between 1637, when the first patent was granted to a woman, and the outbreak of war in 1914, hundreds of women invented unusual and ingenious devices and gadgets, hats and clothes, recipes and tinctures, medical and nursing equipment, toys and educational materials. Huge, profitable companies still exist as a result of the ideas of women such as Melitta Bentz (filter paper for coffee) and Beatrix Potter (merchandising Peter Rabbit) as well as foundations based on the ideas of Maria Montessori (nursery education), Octavia Hill (National Trust) and others. But most women have been forgotten under the assumption that 'there are no women inventors'. Others have invented something highly significant but not been remembered for their innovation. Mary Anderson, following a trip on a New York tram in a blizzard, invented the windscreen wiper. By patenting and registering their ideas women could preserve ownership of them, but it was not until the Married Woman's Property Act of 1882 came into force that they could receive any payment, independently of their husbands, should they go into production. Some, and it has been impossible to ascertain how many, let husbands or employers register their invention for them, thus losing all credit.

These women show that despite all the obstacles in their way, and the issues of ownership of ideas and property, they have always been involved in problem solving: resolving immediate problems affecting their families and themselves, sometimes facing up to issues with a larger perspective. Like men, they enthusiastically embraced the new technologies and fashions of the nineteenth century and tackled some of the many problems of urbanisation.

This, then, is a survey of some of these women; it is not exhaustive as there is no database of women inventors or holders of registered designs and there are many hundreds more hidden away. Beginning in 1637 and ending in 1914, these examples have been drawn, mainly, from the thousands of patent holders and holders of registered designs. The majority are British, although some from the

USA, Canada, Australia, France and Germany who registered or patented their inventions and designs here, or whose ideas were of great significance, are also included. They make a fascinating medley of innovation, eccentricity, sensibility and practicality and show that women have invented all manner of things and would have been as involved as their men in the great technological excitement and changes of the nineteenth century had they been allowed access to the same environments. From them, it is possible to glean an indication of what affected their lives: after all innovation and invention usually arise from a direct need. It is no coincidence that necessity is called 'the mother of invention'.

These women have called themselves inventors, electricians, engineers, metallurgists, milliners, nurses, doctors, authors, publishers, artists, teachers, housemaids, gentlewomen, spinsters, married women, duchesses, ladies and of no occupation. But they have each pursued an idea and registered it and in so doing have given us great insight into their concerns, imagination and determination. This book is a revelation of just some of their ingenious ideas; a full, chronological listing of them appears in the Appendix.

Acknowledgements

The research for and writing of this book have meant pursuing all varieties of leads, some obvious, others obscure. Almost every one has led me to new and unexplored territories. I am grateful to all those who have answered my, at times, vague and naïve enquiries, but your answers have been essential and highly informative.

I would especially like to thank Stephen van Dulken and the staff in 'Science 1 South' (Patents) at the British Library. Their invaluable skills, energy and patience are greatly appreciated. They showed me how to access the patents and trademarks and then retrieved endless volumes. Stephen has shared the journey of this book, from the original idea through to its fruition with the inevitable deviations along the route. His enthusiasm for and understanding of the mind of the inventor are exemplary. No copyright is claimed by patent offices in the text or drawings of a patent specification.

My agent Mandy Little and Jaqueline Mitchell at Sutton have shared my enthusiasm and amazement at some of the things I have found. I am extremely grateful to Sandi Toksvig for writing the foreword. Susan Bennett has been her usual mine of information. Corinne Julius posed the question and made me think. Dr Valerie Mars gave much valued advice on the Victorian domestic scene and Ros Osmond enabled me to gain enough confidence to tackle the 'science ones'. Julia Eccleshare conveyed to me her passion for Beatrix Potter. Nigel Hobden's IT expertise has efficiently introduced me to the bewilderment of a new Apple Mac. Professor Sir Christopher Frayling relayed his interest from afar. Eunice Wilson amazed me with her bundles of names. Fortuitously the subject of Valerie Wickham's enquiry to Stephen van Dulken makes an appearance in my book. The late Maria Tindall was fascinated by this project and I have missed discussing Mrs Coade's stone and Elizabeth Wallace's interior designs with her. Clare Jackson and Helen Holness, at Sutton, have worked hard 'behind the scenes'. Thank you all.

More thanks must go to the numerous museum curators and archivists have been very generous with their time and expertise, especially Rosemary Harden and Elsie Barber at the Museum of Costume in Bath; Edwina Ehrman and Karen Fielder at the Museum of London; Marion Nichols, Alison Taylor and Veronica Main at Luton Museum; John Heyes at the Museum of Childhood in Edinburgh; Terry Berry at the Oldham Local Studies and Archives; Joy Emery of the Patterns Archive, University of Rhode Island; C. Anne Wilson of the Special Collections Department at the Brotherton Library, Leeds University. There are others too, including the librarians and archivists at the Records and Historical Department at the Foreign and Commonwealth Office; Fortnum and Mason company archives; The London Library; Melitta GmbH & Co, Minden, Germany; National Archive of Design at the Victoria and Albert Museum; Public Record Office; Royal Institution; Royal Society of Arts; Margaret Steiff GmbH; Thomas Cook Company Archives; Westminster City Archives and The Women's Library, London Metropolitan University.

In this age of the superhighway, the anonymous web masters of various sites have provided access to essential information easing the pressures on the researcher. Notable among these have been: Institute of Electrical Engineers, Institute of Physics, The Nobel Foundation, Patent Office (UK), Public Record Office (especially the 1901 Census) and the United States Patent and Trademark Office.

Elizabeth Duff has, as always, been a fantastic support. Her comments on my manuscript, advice and amendments have been taken on board. Thanks must also go to Amelia Rowland for efficiently organising the footnotes and bibliography and Thomas Rowland for his thorough reading of the book in preparation of this paperback edition. All errors are now of my own making. Finally, my mother Mavis Jaffé, my husband George Kessler and my daughters Madeleine and Flora Kessler have patiently lived with 'my women' for the last two years; their understanding of my passions and distractions is much appreciated.

CHAPTER ONE

Innovators

Innovation and discovery are crucial to human progress and, since history has been recorded, women have been essential to the development of new ideas and practical solutions to problems. The freewheeling, creative imagination of the inventor knows no bounds. These endless possibilities may be used to resolve a specific problem, to improve a current situation or to invent something new. The results can be unusual, practical, ingenious, obvious, labour-saving and world-changing. They may result in a new gadget, something multifunctional and purposeful or something useless. But they are all fascinating attempts at improving a situation and women's experience and ingenuity have been crucial to their evolution.

During the inventor's paradise of the nineteenth century and the changes wrought on society by the transformation from an agrarian, artisan economy to one of urban, mechanical, mass production, many women were busy, though rarely given any credit for what they achieved. Some even had to give their ideas away to husbands who reaped all the rewards. During the period, inventors sometimes veered towards the fantastic as the study of science became popular and magazines were published to celebrate the new powers of electricity, engines and technologies. In the seventeenth century many women preferred to study the sciences, finding them easier and more accessible than the classics, and by the nineteenth century many were welcome members of the audience at lectures at the Royal Institution, where they would listen to Michael Faraday and others.[1] However, they were unable to sit on any committees or to lecture themselves. Asa Briggs describes the late Victorian period of invention and technology as one where 'There was an element of play, also of entertainment, both in fashion and in invention; and in the discussion of scientific inventions there could be an active element of fantasy. There were many "eccentric ideas."'[2] But it is precisely the 'eccentric' that is vital to the energy and ideas of the inventor.

Maybe as a result of their limited access to the scientific world, women, during the nineteenth century, fought hard struggles to be accepted in medical schools as mathematicians and as discoverers.

Women attended lectures on science which were reported in the *Illustrated London News*. Here, the popular scientist 'Professor' Popper demonstrates his induction coil which could produce a spark, 'or flash of lightning, 29 inches long'. (*Illustrated London News*)

Retaining ownership of an idea is very important to its originator, especially if it is a concept with commercial or practical implications. Legally this can be done by patenting that idea, but only if it is new, inventive and has a wider industrial application which gives the holder ownership of the idea for a number of years. The granting of a patent, therefore, is official acknowledgement of ownership of that idea. Designs are registered as protection of their copyright, again acknowledging their ownership by the originator. It was during the Tudor period that patents were first introduced and Elizabeth I granted about thirty, most for the manufacturing process of soap, saltpetre, alum, leather, salt, glass, knives, iron and paper. Over the centuries patent legislation, especially after the Great Exhibition of 1851, has been refined and reformed to deal with a changing society. Today most patents are for improvements to technological processes. The introduction of the Designing and Printing of Linen Act of 1787 was the beginning of the copyright protection for designers and, again, the Act has been changed and modified to deal with new and different design processes and technologies.[3]

Patents are a fascinating study not only for the ideas and innovation they are introducing but also for the background details they give of the inventor, the times in which they lived and reasons for their improvement. Many are accompanied by detailed diagrams and drawings which are also worthy of study. However, the granting of a patent does not necessarily mean the object went into production; it is simply acknowledgement of ownership of the idea.[4] The same applies with a registered design. Patent applications give historical insight into why something could be improved upon, the effects of a new material or technology and the aspirations of the inventor. Within eighty years of Elizabeth I granting the first patent in 1561, Amye Everard als Ball would become the first woman to apply for one. She was a widow, who asserted that her preparation for a 'Tincture of Saffron, Roses &c' was her own invention. Ever since, there has been a slowly increasing number of women, albeit far fewer than men, who have taken out patents to protect ownership of their inventions and intellectual property. Between 1617 and 1852 only sixty-two patents were awarded to women in England. From 1852, when a new patent act was introduced in line with the growing industrialisation and to incorporate England, Wales, Scotland and Ireland, patents in women's names, rose to about 2 per cent a year. (In 1898, 638 patent applications were made by women in Britain against a total number of 27,639.)[5]

Sometimes the ability to make and design items led to women not only patenting their designs but gaining greater confidence and independence by doing so. As with the straw-hat makers, many of whom, like Jane Lowrey, patented their improvements to straw plaiting. The wearing of a straw bonnet made from Leghorn straw plait was the height of fashion in the 1800s and resulted in a huge hat-making industry in Luton, where its women straw plaiters earned more than their husbands and became a highly independent and vociferous group. Madame Roxey Caplin ran a corset shop in Berners Street, London and with her husband patented their inventions to stretch the neck and back. In 1864 she published her book on *Health and Beauty; or woman and her clothing*, based on the exercises devised in her gymnasium and her designs for corsets to improve posture. Eventually the corset was modified and by the 1890s the first brassieres appeared, again designed by women.

The introduction of a new technology could have enormous impact on women's lives and inevitably they produced innovative methods to cope with it. When the bicycle arrived in the 1880s they designed numerous devices for a woman to hide her ankles, keep her hat on and protect her clothing, as well as a divided skirt enabling her to sit, modestly, astride a saddle. Others used new technologies for philanthropic purposes. Amelia Lewis, 'trading as Mrs Amelia Freund', patented numerous stoves that were small, easily portable and fuelled by peat to enable the poor to keep warm and eat nourishing food: she was a great advocate of steaming food and temperance living. Lady Mary Gladstone patented her huge and luxurious travelling trunk in 1890, clearly someone with sights on travelling the Empire. No doubt aware of the skirmishes within the Empire the inventive nurse appears from the 1880s with new seating devices, stretchers and methods of keeping a bandage in place.

Some women became administrators of their deceased husbands' estates and it is unclear whether they had any involvement with the design of the initial invention. Others, such as Mrs Burgess who inherited the firm of John Burgess and Co., makers of anchovy essence, with her stepson, took their deceased husband's places and ran the companies. Records show Mrs Burgess's involvement in the running of it, her patent for preserving anchovies and the inevitable rifts that developed between her and the stepson.

There are the endearing and – what Asa Briggs refers to as – the 'eccentric' inventors like Clara Louisa Wells who lived abroad and whose schemes included an 'Improvement in Aerial Locomotion': a raised railway system around the Bay of Naples, dependent, in part, upon suspension from large balloons. In 1903 Baroness Margarethe Johanne Christianne Marie von Heyden, from Germany, patented 'A New or Improved means for Preventing Coition in the Case of Bitches and other Female Animals': a canine chastity belt. There are numerous patents from German women after the country was unified in 1871 and the majority are for engineering or mechanical applications. The diagrammatic drawings for these are very detailed, reflecting Germany's growing engineering strength and a more enlightened attitude towards women's education.

Having decided upon a purpose for their inventions, many women were delighted when they could fulfil more than one function, as with Mrs Florence Crosby Hurd's 'Appliances for Inducing Correct Breathing through the Nose instead of the Open Mouth; for Flattening Prominent Ears and Keeping Surgical Bandages and the like in place on the Head'. By 1904 there was an increase in patents for items of a personal nature: that colonic irrigation was a peculiar trait of not only the 1990s can be seen by the numerous applications, by women, for equipment relating to enemas and douches. Miss Elizabeth Keswick of Ilkley and her 'Improvements in or relating to Apparatus for use in connection with Enemas, or Irrigators, and the like Injectors' was one of many.

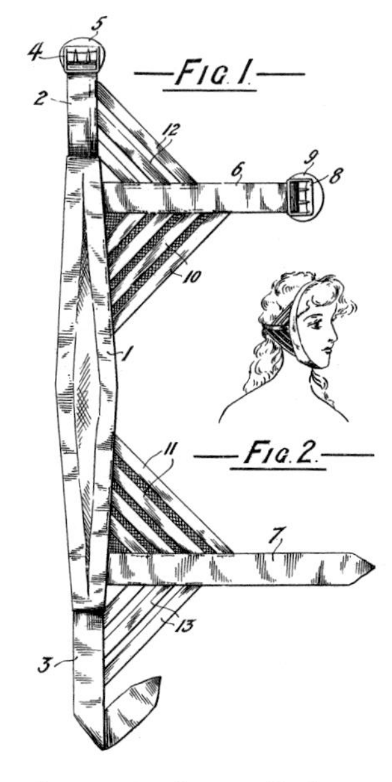

Florence Crosby Hurd's device to improve breathing, flatten the ears and retain head bandages in position. Patent.

CHAPTER TWO

Early Ladies

When the widowed Mrs Ball was 'Preparing a Tincture of Saffron, Roses &c',[1] England was a very different country from the noise and bustle of industrialisation that would dominate the landscape in the nineteenth century. In 1637, Charles I was halfway through a reign which witnessed terrific civil unrest and eventual war in 1642, as a response to his dissolving of Parliament and belief that he was answerable only to God. Eventually the people's disillusionment in their King was so great that he was beheaded outside the Banqueting House in Whitehall in 1649. Scotland was not yet part of the union[2] and England was intent on colonising the east coast of North America.

Before the advent of the modern chemical processes, saffron was used for dyeing, in the cooking of food for the wealthy, as a medicinal remedy and in the preparation of perfumes. Mrs Ball's patent to preserve and extend the life of saffron would have been invaluable. Saffron, which was then, as now, very expensive, derives from the stigmas of the crocus flower, and Saffron Walden in Essex was a cultivation centre for it. She describes her method to preserve, at maximum strength, the expensive, rich, golden yellow saffron as her own invention:

> shall dissolve into tincture, and gaine out of everie pound of saffron soe made vp severall ounces, which shall continue and remaine in full strength and virtue for manie yeares more than saffron in the sheyve or leafe vsually doth or can.

Amy Potter's patent in 1678 was as a direct response to government legislation that demanded the dead be buried in woollen fabric, as a way of supporting the woollen industries. Hers was

> An invention for makeing of Flanders Colbettine and all other laces of woollen to be vsed in and about the adorning or makeing-vpp of dresses and other things for the decent buriall of the dead or otherwise

> which we are informed will be to the increase of the woollen manufacture, and encourage obedience to a late Act of Parliament, entitled an Act for Burying in Woollen current act of Parliament for the interment of bodies in wool to protect the woollen merchants with her invention.[3]

Probably, precisely because they were widows and no longer wives, Mrs Ball and Mrs Potter were able to claim ownership of their ideas. A wife was still to be regarded as a husband's appendage until 1870. This is clear in Thomas Masters's patent in 1715.[4] Thomas Masters, a planter and merchant from Pennsylvania, still an English colony, lodged a patent application in England, clearly stating that it was on behalf of his wife Sybilla.[5] The following year Sybilla Masters herself did manage to patent her second invention in her own name, this time for a type of straw plait from which to make bonnets.[6] The Masters's patents are interesting as they reflect the competition among straw-bonnet makers in Europe and North America, all eager to produce a cheap type of straw or plait which could emulate the expensive lustre of the type produced in Leghorn, Italy. One hundred years later this would become the basis for the huge hat-making industries in towns such as Luton in Bedfordshire. The Society of Arts in London sent seeds to America to be grown and the harvested straw was brought back to be plaited in England. Many of these early American patents and inventions reflect the settlers' lives of developing the land and growing crops, where women were as vital to survival as men. In her fascinating book of American women inventors, *Mothers and Daughters of Invention*, Autumn Stanley[7] highlights the numerous implements invented by American women to cultivate the land, grow and harvest crops, as well as the tools for homemaking.

By the late eighteenth century, women in England were becoming more vociferous, in what Linda Colley refers to as the period when they first began to be regarded as citizens, seeing themselves as part of the wider society.[8] In 1792 Mary Wollstonecraft wrote her seminal book *A Vindication of the Rights of Women*. Her own life was a complex mixture of determination and romanticism, as she promoted her ideas about women's hidden abilities and the appropriate rights they should be allowed. She herself was enamoured of the artist Henry Fuseli; had an illegitimate child, Fanny, by the American, Gilbert Imlay; travelled abroad with her husband, William Godwin; and finally died in childbirth with their daughter, Mary. Mary Godwin, who married the poet Percy Shelley, has continued to entrance generations with her creation: *Frankenstein*.[9] At the same time Georgiana, Duchess of Devonshire, was openly influencing political events at Westminster.[10] Her extravagant lifestyle and love of

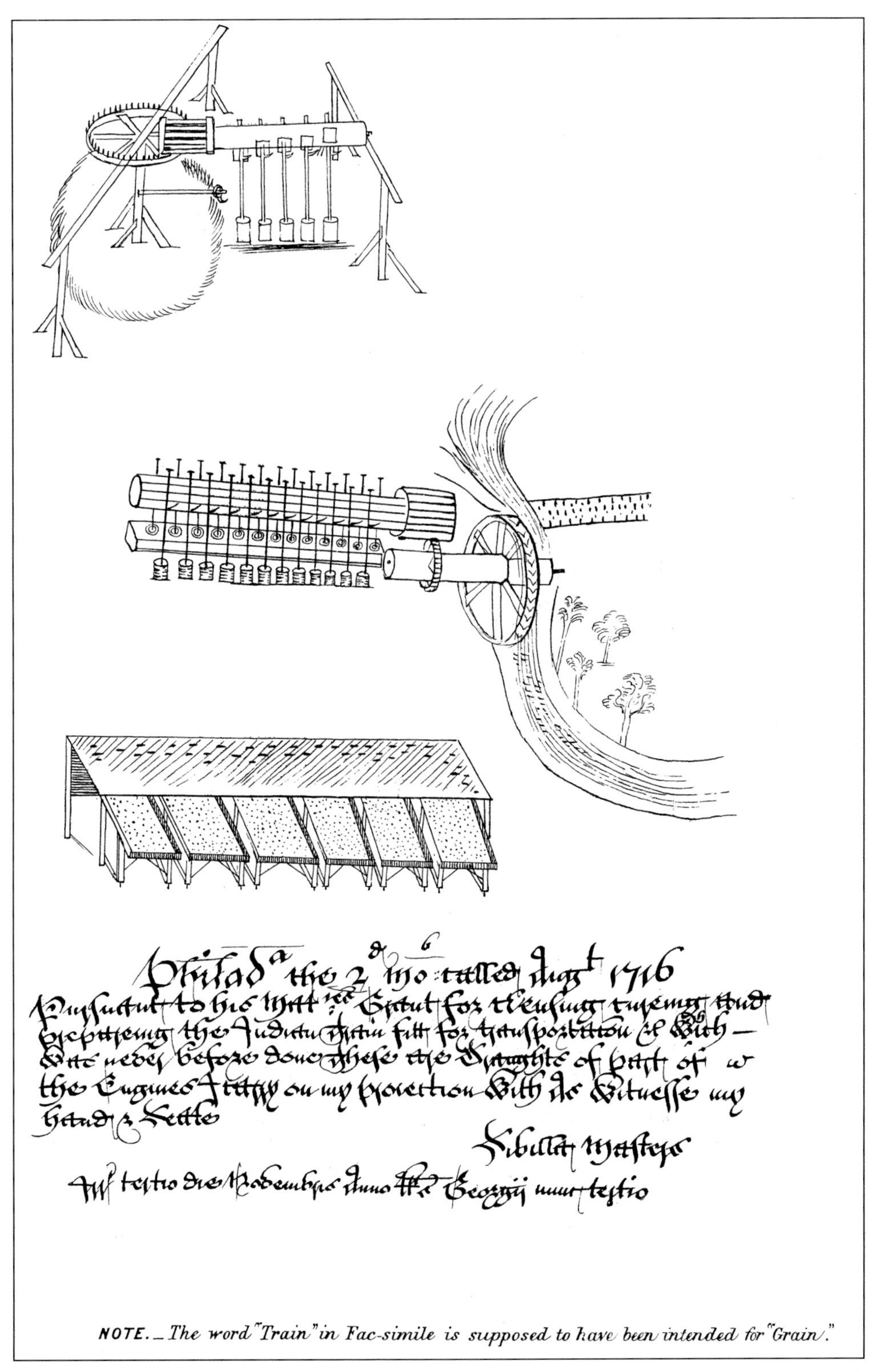

The diagram accompanying Thomas Masters's patent application, to clean and cure corn, in which his wife Sybilla's name appears on the bottom right, accrediting the invention to her. Patent.

Examples of Mrs Coade's stone surrounding the doorways designed by John Johnson in 1778, at 61 New Cavendish Street, central London. (*Author's Collection*)

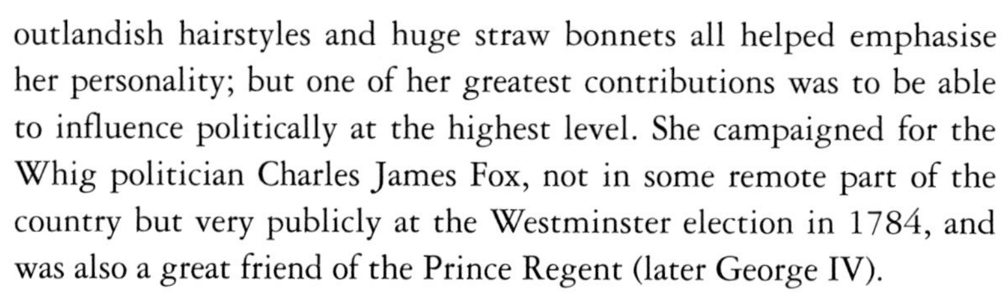

outlandish hairstyles and huge straw bonnets all helped emphasise her personality; but one of her greatest contributions was to be able to influence politically at the highest level. She campaigned for the Whig politician Charles James Fox, not in some remote part of the country but very publicly at the Westminster election in 1784, and was also a great friend of the Prince Regent (later George IV).

Meanwhile, across the river in Lambeth, Mrs Eleanor Coade was running a factory making her unique Coadestone. The Coade family originated in Cornwall where Eleanor's father had become bankrupt. The law did not permit his wife to handle any money, so what was left of his factory was handed to his daughter Eleanor (known as Mrs Coade). It was she who, in 1769, moved the factory to Lambeth in south London. Most natural stone does not weather well and Mrs Coade hit upon a secret formula to make a ceramic-based, hard-wearing material, much of which has stood the test of time and still adorns many façades in London and towns around the country. Not only did she manufacture her artificial stone but she also employed a host of designers and craftspeople to design and make the moulds. Elegant and intricate designs for Grecian-style urns, figurines, leaves, flowers, fossils and animals, were made to adorn friezes, pediments and pillars of the buildings by leading Georgian architects, such as Sir John Soane and Robert Adam. The doorways in Bedford Square and the friezes on Schomberg House in Pall Mall in London are all of Coadestone and the Museum of London has fine examples on display. So popular was Mrs Coade's work that she produced a catalogue in 1784 to show all her latest designs.[11]

The demand for corsets to be more comfortable, as well as addressing concerns about bodily health and safety, was already apparent in 1800. They were made on an individual basis for the wealthy, and even Georgiana, Duchess of Devonshire, designed her own.[12] Maids often received their mistresses cast-offs. The dressmaker, often a woman, was crucial to these developments. Martha Gibbon, a dressmaker in Covent Garden, designed 'A Certain new stay for Women and Others'.[13] She was clearly worried about the effects on the bowels of tight lacings, so her stays (stays were to the eighteenth century what corsets were to the nineteenth and twentieth) were made in two parts: a front and a back, fastened together at the sides with laces and metal so that they might expand or contract depending on the state of the wearer and any 'sudden pressure on the bowels' so that 'the abdomen is supported, and the bowels prevented from passing through the momentum, which their gravity will frequently do for want of such support and sometime occasion the hernia or rupture in the groin.'

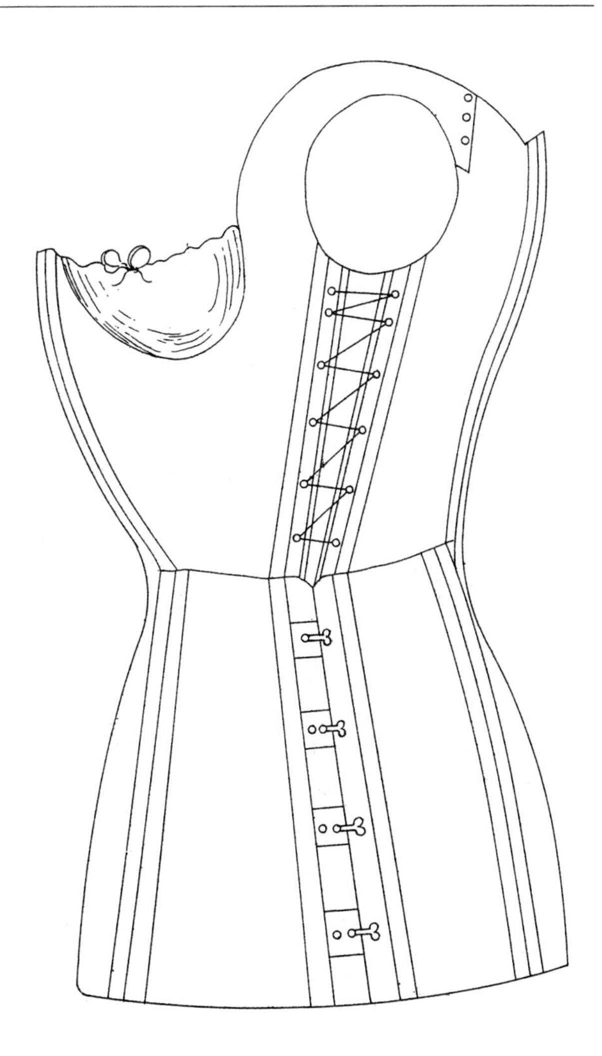

Side fastening stays that could be loosened or tightened as necessary, designed by Martha Gibbon. Patent.

Bodily functions were further considered when Mrs Ching developed the patent taken out by her late husband in 1796. Worms were widespread at the time and in 1808 her 'Worm Destroying Lozenges'[14] appeared, which she recommended taking night and morning to purge the body of them.

These women, and many others whose achievements have gone unrecorded, were important in the movement to cultivate women's independence of thought and achievement, which 100 years later would express itself in the suffrage movements in Britain and North America. But the road was inevitably hard as women – as well as the majority of men – were still without the right to vote. Married women were denied ownership of property independently of their husbands and, no matter what the reason for a divorce, lost all access to their children. Many had to hand over ownership of their ideas to their husbands. Yet, despite the obstacles, their creativity and imagination, like that of any inventor, did find expression. It was in the nineteenth century that women's ingenuity, like men's, showed

A group of suffragettes during the 1910 General Election. (*Illustrated London News*)

their enthusiasm and determination in the quantity and variety of inventions they devised. Eventually, in 1870, the Married Woman's Property Act enabled them to keep £200 of their annual earnings; in 1877 the Act was extended to include Scotland and in 1882 brought in the ownership of property. Finally, in 1884 they were regarded as individuals independent of their husbands.

CHAPTER THREE

Appearances: Bonnets, Corsets and Umbrellas

It is not surprising that matters relating to their own bodies provided women with much scope for industry. Not only do clothing styles change at the whim of fashion but also through the influences of new fabrics and technologies, and by the demands of women's attitudes to themselves and their lives. Their inventions were sometimes motivated by the supposed health improvements offered by a new corset or the use of a new material, like elastic. In the 1820s, many people, including Thomas Hancock from Islington, were finding ways to use rubber; at first he made rubber cuffs for gloves which he developed into strips and then thread from which elastic was woven.[1] The application of elastic to clothing had a profound effect: it transformed fastenings, stockings and corsetry. Whether by means of an elastic side to a shoe or corset, a new type of hatpin or a divided skirt, women's concerns to be more comfortable, elegant, healthy, noticeable or assertive of their newfound independence were illustrated in fashion. Some of these inventions were very practical, others decorative and alluring. Their manufacture often also involved women – the hat-maker, seamstress and dressmaker – who were precursors of the modern fashion designer. Women have been vital to the success of the corset making, elastic, shoe and tailoring industries.

Hats and Bonnets

Straw woven into plait and made into bonnets and baskets has a great and diverse tradition. Many cultures have produced their own distinctive styles of plait dependent upon the type of grass grown, the colour, texture and lustre of the straw it produces and methods of weaving. Paper strip, slivers of wood, leaves, silk and wool have all been woven with the straw into the plait to create an even more exotic and unusual result.

When, in 1715, Thomas Masters from Philadelphia patented his wife Sybilla's invention in his name, it was their intention to market corn, which they named Tuscarora rice, in England. Undeterred by its poor reception, Sybilla Masters applied for a second patent, this time in her own name, in 1716,[2] in which it states,

> Sybilla Masters, wife of Thomas Masters, of Pennsylvania, Merchant, hath humbly represented vnto vs by her peticon, that she hath with great charge, paines and expence, invented and brought to perfeccon 'A NEW WAY OF WORKING AND STAINING IN STRAW, AND THE PLATT AND LEAF OF THE PALMETA TREE, AND COVERING AND ADORNING HATTS AND BONNETTS IN SUCH A MANNER AS WAS NEVER BEFORE DONE OR PRACTISED IN ENGLAND OR ANY OF OUR PLANTACONS.'

This was for a straw and leaf woven material, to produce plait for hats. The plait would then be neatly stitched, side-to-side, to form the bonnet or hat.

In the eighteenth century, fine silk was often woven into the plait in an attempt to make the indigenous straw emulate the sheen of that produced in Leghorn (Livorno) in Italy. Leghorn straw plait was expensive and highly sought after by wealthy, fashion-conscious women in England and France. The grass from this area was grown for the rich, golden lustre its straw produced and was an important export for the area. In order to plait straw it has to be split to make it more pliable and most grass will then produce one side with sheen and the other matt. Leghorn did not have to be split and thus retained its universal lustre, making it even more special and expensive than the local, English straw. One Leghorn bonnet would be purchased and adaptations made to it each year according to the dictates of the latest fashionable style. The desire and pressure to produce a cheaper straw plait with a lustre equal to that of Leghorn straw was enormous. To achieve a method of emulating or improving upon Leghorn would have been, commercially, highly lucrative. Women and men, like Sybilla and Thomas Masters, in both Britain and North America, were desperate to find new grass seeds or a method of plaiting.

The Duchess of Polignac (*c.* 1749–93) by Louise Elizabeth Vigée-Lebrun. She is wearing a straw bonnet probably made of Leghorn straw plait. (*Private Collection/Bridgeman Art Library*)

In 1809, Mary Kies[3] became the first woman to be granted a patent in the United States. She devised a method of weaving silk and straw, which greatly enhanced the fledgling American millinery industry. Grace Elizabeth Service took out her English patent in 1815[4] for 'A New method of Manufacture of Straw with Gauze, Nett Webb and other similar articles, for the purpose of making into hats, bonnets, workboxes, workbags, toilet boxes and other articles'. Both these women were trying, by weaving silk and gauze into the straw, to make it look as rich and shiny as Leghorn plait.

The need for a cheaper, home-produced, Leghorn-like straw dramatically increased as a result of the trade blockades across Europe, caused by the Napoleonic Wars (1799–1815). The grasses of the rolling countryside of Hertfordshire and Bedfordshire helped to establish Luton as a centre for the straw plait and hat industry, which brought work to hundreds of women. Once the wars were over it was vital to preserve this new industry, which was capable of producing a good substitute for its rival, and avert a flood of Leghorn straw arriving on the English market. From the beginning women were instrumental in finding ways to emulate Leghorn and the consequent success of the hat-making industry in Luton.[5] William Cobbett, having been impressed by the straw bonnet made for him by Sophia Woodhouse (a captain's wife whom he met on one of his many voyages), recognised their contribution and used the Luton straw plaiters as an example in his book *Cottage Economy*. In 1824, the Society for the Encouragement of the Arts, Manufactures and Commerce[6] acknowledged the importance of the industry[7] and the need to preserve its jobs, and rewarded them for their production of Leghorn-type straw. Lucy Hollowell of Banbury received 15 guineas for her method of using seeds of the *Poa pratensis* grass to produce an improved straw.[8] The seeds had been sent by Mrs Wells, the daughter of a farmer in Connecticut. In 1825 they further acknowledged her work and that of Sophie and Anne Dyer of Alton, Hampshire; Mrs Venn of Hadleigh, Suffolk; and the children of Mrs Villebois's school in Berkshire for their 'bonnets and hats made of British materials, platted and knit in imitation of those imported from Leghorn'. They noted the craftswomen's manual dexterity and ability to reproduce a straw with a similar lustre to that from Leghorn. Mrs Jane Bentley Lowrey, an Exeter hat-maker, not only received the Society of Arts Ceres Medal in 1825, for her method of plaiting two pieces of straw together to reproduce the lustre, but also patented her method in 1829.[9]

Although the women hat-makers of Luton did not have to face the appalling, dirty and dangerous factory conditions of women and children in the Lancashire and Yorkshire cotton and woollen trades, the work was undoubtedly hard. This was a seasonal trade, paid on a piecework basis; each length of straw had to be licked to keep it moist, causing terrible sores on the lips. Like the lace-makers from Honiton and the knitters and weavers of the Scottish Islands, they worked in their own homes or in small groups with their children, the older ones also employed as workers. Not only did they produce the varieties of straw plait essential for the industry but were instrumental in stitching it into hats and bonnets and the final decoration. They became a formidable and refined group, practising their craft with pride. Their straw hats were sold at Cloth Fair, the hat trade area of the City of London. A charabanc full of these Luton women made a spectacular entrance into Hyde Park at the Great Exhibition, and they were, no doubt, as enthralled as the majority of visitors, including Queen Victoria, by the novelty of the Crystal Palace and its contents.

Top: Dealers buying straw plait from Luton women, 1885. (*Luton Museum Service*) *Above*: The interior of a Luton hat factory, 1906. (*Luton Museum Service*)

As the century progressed, factories, more akin to extended workshops, grew, employing local women. With the passing of the Married Woman's Property Acts between 1870 and 1884, they were often able to earn more than their husbands, making them some of the highest paid hat-workers in Britain with a consequent independence and self-confidence. In the 1890s a 'girl machine hand, took £3 5*s* 8½*d* in her highest week, in her lowest week 2*s* 4*d* and in the whole year she earned £62 7*s* 7*d*.[10] This is the most that any girl could earn. Their lengths of straw plait were sold at Luton market to hat-makers and these women extended their abilities beyond being plaiters and machinists to become designers and managers. In 1914 between 5 and 10 per cent of the workshops and factories in Luton were run by women.

Although straw plaiting remained important, the industry was also using other methods for hat-making such as felt moulded on blocks and decorated. Mrs Durst-Wilde, was a widow who lived in Paris, and probably came from a firm of milliners, as her husband and sons also took out *brevets* (patents). She brought a patent application to London on behalf of C. Valentin Chevy Jun.,[11] in which she described enhancing the sheen of hats with the application of metallic lustre of gold, silver or bronze metallic powder. The same year she also applied for a British patent,[12] in her own right, for 'Producing Raised Patterns on Hats, Caps and Bonnets' that related directly to hats of felt or other material that would not be damaged by damp conditions. The hat was stretched over a block, then a second, relief block which had designs and patterns embossed in it was placed over the first to create a textured, raised surface. At the same time in France she took out two more *brevets*.[13]

Women's growing emancipation and the arrival of the bicycle were reflected in the change in hat fashions. By the 1890s, as the age of the straw bonnet declined, there was an increase in the need for waterproof materials and hats that would stay in place, with or without hatpins, while cycling. A plethora of inventions emerged to meet these new demands.

Corsetry

The progress from the stays of the seventeenth century via the corset and suspenders, to the modern bra and tights does, at times, appear to be akin to a history of torture as women pulled and pushed their bodies into constricting garments, occasionally eased by a new type of boning, steel ribs or elastic. Patent applications made by women, reveal overlooked aspects of concern and design and show variations from those made by men. There was also an assumption of preventive medical benefits from wearing a fitted corset as well as the aim to achieve something more comfortable but at the same time elegant.

The Covent Garden dressmaker Martha Gibbon described, in terrifying detail, the potential side effects of not wearing one of her new corsets. Not only would hers prevent hernias, troublesome bowels, ruptured groins, it would also apply gentle pressure to the 'upper belly' to act as a truss thereby preventing 'any protuberance of the intestine'. Padding could be applied in the appropriate part for 'persons to whom nature has not been

A Little Tighter by Thomas Rowlandson, published by S.W. Fores in 1791. (*Private Collection/ Bridgeman Art Library*)

favourable, [to] combine ease, grace and elegance . . . the symmetry of the female shape. . . . [and] remove all external appearances of crookedness and deformity'. By the 1840s Mme Roxey Caplin and her husband were focusing on the relationship between good posture and health, making use of his medical knowledge and her ingenuity. In 1905 Mrs Maria Vermeulen from Brussels devised her 'Abdominal Belt for Enceinte or Lying-in Women and Abdominal Sufferers combined with Bust-supporter and Waist Band'.

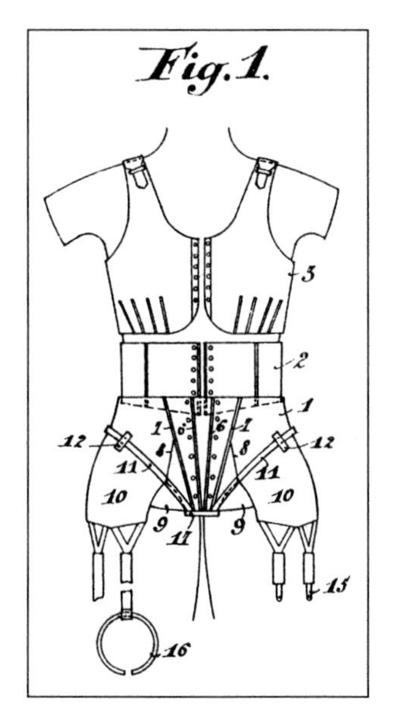

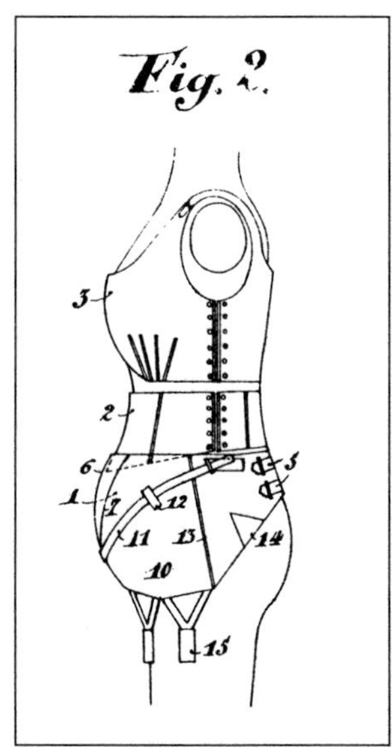

Maria Vermeulen's corset for a pregnant woman. Patent.

Before mass manufacture, corsets were very heavy and thick, with two layers of fabric, an inner one of cotton and an outer of coloured silk or brocade with whalebone ribs sandwiched between the two. Lacing could be along the back seam or at the sides with additional metal fasteners down the front, the top might be edged with lace. They were worn over a thin cotton chemise to protect them from the body's perspiration and under fashionably shaped petticoats or crinolines and the dress. With their tight lacing and stud fastened fronts, whaleboning and later steel ribbing, they constricted the wearer's posture, making the waists as small as possible, pushing up and supporting the breasts and enforcing the straightest of backs. Until the end of the nineteenth century they were handmade, by a corsetière, for upper-class socialites and the growing market of middle-class women. Gradually male corsetirès were replaced by women, who, like Mrs Gibbon, set themselves up in business, not only as personal fitters but also as designers and makers. They introduced a personal dimension to corset making based on experience and knowledge of a woman's shape which could challenge the traditional male view of an ideal shape.

It has been assumed that working-class women did not wear corsets but Leigh Summer[14] has found evidence that they often receive their employers' cast-off garments, only to be accused in some cases of looking like women above their class. With the advent of mass production in the 1890s, corsets became available for all, although they were by then less ornate and usually made out of calico.

The role of the corsetière cannot be underestimated. Increasingly women ran these businesses. In 1856 Mesdames Marion and Maitland advertised their corsets in the *Lady's Newspaper*. With premises based in Oxford Street and Connaught Terrace, near Marble Arch, they fitted their customers in Marion's Resilient Bodice or their Corsaletto di Medici that was described as 'having resilience in conformity with the movements of respiration'. But it was probably Mme Roxey Caplin who, with her husband Dr Isidore Caplin, went furthest in designing corsets for every shape, size or physical state as well as exploring the relationship between good posture and health. Mme Caplin had a shop at 58 Berners Street, off Oxford Street, and another at 55 Princes Street, Manchester. Two patents were taken out in Dr Caplin's name, which were probably devised by both of them.[15] The second was a frame to support the back, straighten the neck and hold the head high. It was Roxey, as a 'Manufacturer, Designer and Inventor', who was awarded the Prize Medal at the Great Exhibition in 1851 for her 'self adjusting corsets, child's bodices, ladies' belts'. Not only did Mme Caplin have her shops but also a gymnasium where she advocated the importance of exercise. In her book, *Health and Beauty: or, Woman and Her Clothing, considered in relation to The Physiological Laws of the Human Body*[16] she wrote copiously and respectfully of her appreciation and understanding of the body and how best to preserve it:

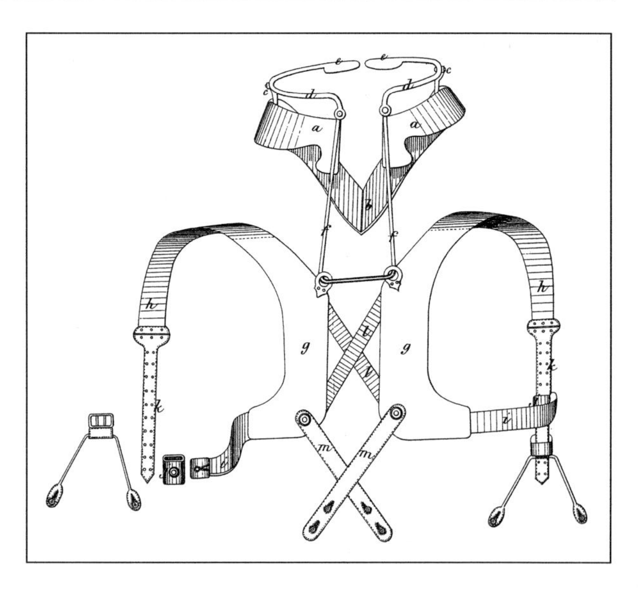

The self-adjusting frame designed to support the back and straighten the neck devised by Roxey and Isidore Caplin in 1852. Patent.

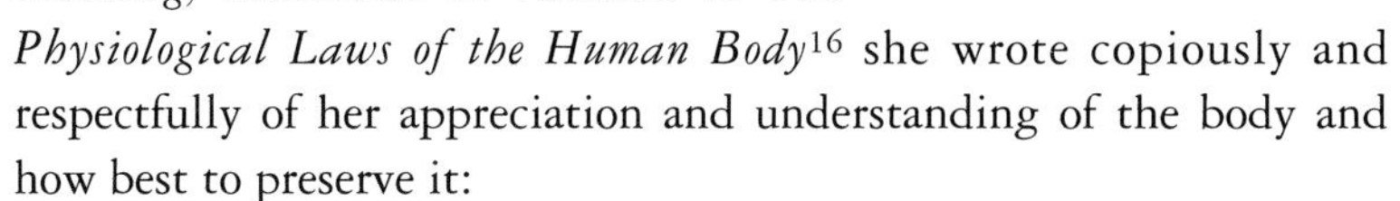

> The mechanical powers, the geometrical figures, the motion and weight of the fluids, and the operations of chemistry, are continually engaged in the support and renewal of the frame; uniting an accumulation of force with a simplicity of operation truly wonderful – all contributing by their million operations to produce that beauty of the outward form which we are now seeking to illustrate. See how the head – the seat and treasure – chamber of their soul – is balanced on the neck; the spine, how it curves in beautiful arches; the body, with its graceful lines, enclosing, as in a casket, the other vital organs. We must come to regard this mass as a whole, and must, as before remarked, understand the utility and relations of the various organs to one great end, before we can properly appreciate the conditions of health and the means of cultivating or preserving beauty.

Hers was an ergonomic approach to the design of corsets whereby aspects of posture and health in relation to the workings of the body were incorporated with every possibility of need. She produced more than twenty-four different designs including: the Elastic Front Corset, for ladies who sang; the Corporiform Corset, with invisible props for the overweight; an invisible Spinal Corset for those with a distorted spine; and various medical support belts.

Acknowledgement of a woman's pregnant shape was rarely admitted. Queen Victoria throughout her nine pregnancies had gradually loosened her stays to disguise her expanding shape. In her letters and diaries she described her disapproval of her daughters and granddaughters revealing their fecund shape.[17] But Roxey Caplin in 1864 had designed the 'Self-regulating Gestation Corset', which included elastic in its structure to give support without too much pressure and thus avoid 'abortion resulting from a deficiency of muscular power'. There were other attempts at comfort like the basque and corset with tabs, button holes and straps to enlarge it designed by Elizabeth and Matilda Abbott.[18] Charlotte Smith from Bedford also looked at the necessity to expand a corset with her stays which had an opening down the front and fastenings and straps that could be expanded.[19]

Support in pregnancy was further developed by Maria Vermeulen when she patented her 'Abdominal Belt for Enceinte or Lying-in Women and Abdominal Sufferers combined with Bust-supporter and Waist Band'.[20] The detailed diagram of the garment clearly shows the outline of a pregnant belly and the material is 'shaped to compel the middle part of the belt to bulge or arch outwards'. There was a slit in the fabric behind which a 'piece of pleated fabric is secured with its opposing edges to the belt-halves and arranged upon elastic bands likewise attached with their ends to said belt-halves to keep the said fabric taut and adjusted to the abdomen when said belt is lengthened or shortened'. Around the bottom of each leg of the corset are ribbons each 'carrying to their loop-shaped extensions buckles securable below the knees' as a type of suspender for stockings. These replaced the garter, which was now known to be bad for circulation.

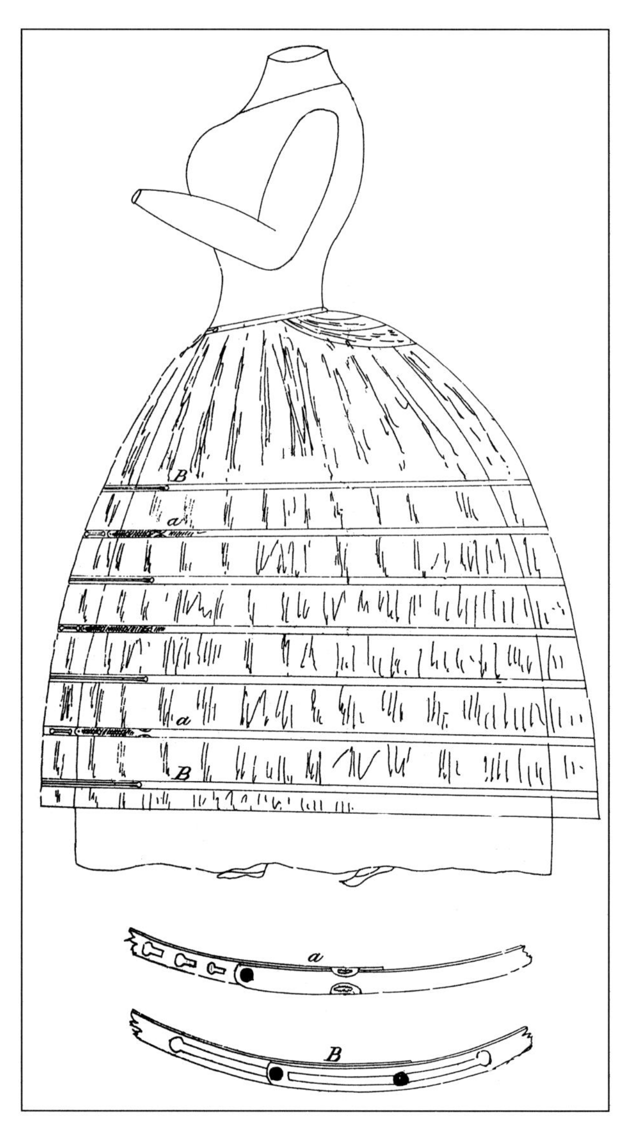

Mary Randle's petticoat to wear beneath a crinoline in which spring hoops were fastened and adjusted in accordance with the width of the skirt. Patent.

Fashion's need dictates female shape and by the mid-nineteenth century this was reflected in a more curvaceous line, coinciding with women's greater self-confidence.[21] The crinoline with its exaggerated width at the hips was highly fashionable in the late 1850s and early 1860s. Its large wooden frame, emphasising the narrowness of the waist, required specific corseting and calico petticoats like that attempted by Mary Randle[22] in 1861. By 1863 a milliner and crinoline manufacturer, Elizabeth Collier of Leicester, acknowledged the inconvenience of wearing crinolines by making them 'capable of contraction or expansion at the will of the wearer'.[23] Her frame hoops were made of steel with joints and hinges so that each one could be lengthened or shortened to alter the volume of the crinoline.

The first sewing machines began to appear in 1851 when the American Isaac Singer patented his machine, having won a legal battle with Elias Howe. Howe had invented a needle and lock stitch machine for home use, whose early problems had been resolved by his wife Elizabeth when she observed that the eye of the needle should be at the base point and not at the top.[24] By the 1860s the Singer Co. had become the world's largest manufacturer of sewing machines. This innovation radically changed approaches to sewing both at home and in factories. In the field of corsetry alone women were increasingly involved in the production of factory-made, mass-produced corsets, making them more widely available to women of all classes. Inevitably, women in the corset factories were poorly paid in arduous and dirty jobs – cutting whalebone and steel, and stitching heavy cloths[25] – the artisan's skill of the corsetière remaining solely for the wealthy. These mass-produced corsets were generally made of calico and were much more flexible than the handmade varieties. Those issued to female prisoners, although of standard and not individual sizes, were specially designed to be looser to allow the wearer to bend easily for work. This need to be more flexible, understood in great detail by Roxey Caplin, that had been acknowledged by the new professional middle class of women in the 1880s who often cycled to work, is reflected in the design of their corsets. Boning no longer covered the hips, and whalebone was replaced by steel either in strip form or spiralled. Suspenders for stockings, instead of garters, were added and trademarks appeared with names like CB Spiral Steels and Prima Donna.

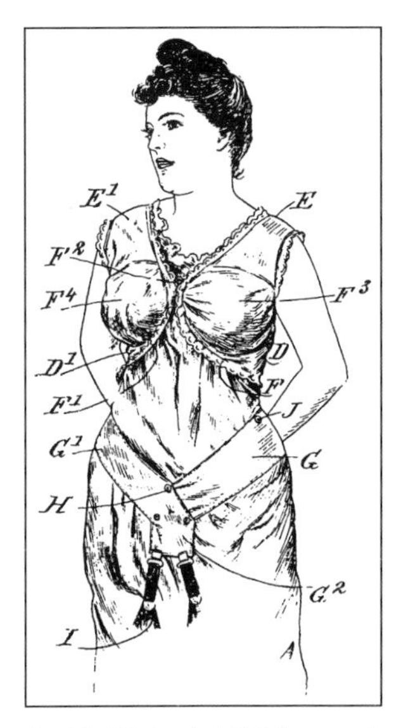

Sophie Vischer's 1903 bust and abdomen supporter as an alternative to the traditional constricting corset. Patent.

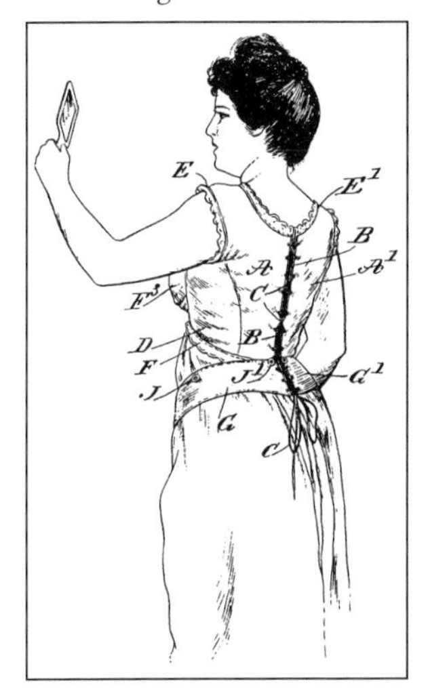

To meet the requirements of a now much more independent woman, the whole concept of underwear began to change and evolve into two distinct garments: the corset/suspender belt and the bra. In 1903, Sophie Vischer of New York City invented her 'Improved Body Garment or Support for the Bust and Abdomen and also for the Skirts and Hose' to replace the traditional corset and release the body from unwanted pressure.[26] It was

> a supporter for the bust, abdomen and skirts . . . arranged to be supported solely from the shoulder bones and hip bones of the wearer, to leave the lungs, stomach and other vital organs completely free of pressure to prevent disfiguration of the wearer's body and the well-known ill effects

> incident to the wearing of tight fitting corsets, to strengthen the back of the wearer, and to provide means for supporting the skirts, hose and the like.

As before, the garment was to be worn over a thin chemise and while the back and bust are supported, the stomach is free from pressure.

Marie Tucek devised a brassiere as distinct from a corset in New York in 1893.[27] What made this garment different was that the breasts were not pushed upwards from the corset but supported and held in place by shoulder straps. She stated that it would be ideal for the new, high-waisted, Empire line dress, fashionable at that time.[28] The brassiere was further developed by Lydia Martell, a married woman from Seattle and Edward Payne from the same address who, in 1912, devised a somewhat bizarre bust supporter, which was part bra and part corset.[29] Where corsets had stressed the importance of posture, pressure on the abdomen and straining of muscles, this concentrated on the support of the breasts, no matter what their size. It consisted of a narrow, waist corset with two hemispheres attached, made of fabric over a wire frame. It was designed particularly for 'women having heavy busts as well as for the use of women having little or no bust. . . . is especially designed to hold up and support the bust, comfortably and properly, in a correct and natural position'.

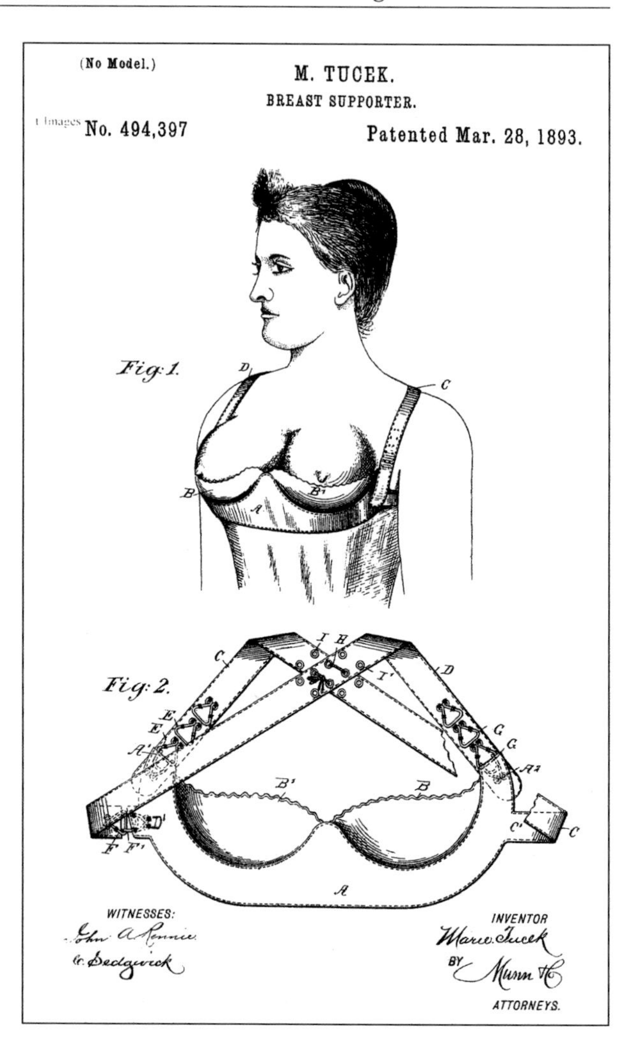

The breast supporter designed by Marie Tucek. Patent.

Suspenders displaced garters as the preferred method of wearing stockings. They were attached to the corset and worn over the long leg length of the knickers or drawers. Bloomers, the long-legged billowing knickers, named after the American suffragist and dress reformer Amelia Bloomer in the 1850s, did not really catch on until the bicycle became popular thirty years later; until then women did not wear knickers. However, when they finally were deemed necessary they brought yet another series of problems to be overcome. Stocking tops had to be hidden beneath the hem of the knicker leg. A drapery buyer, Emily White, from

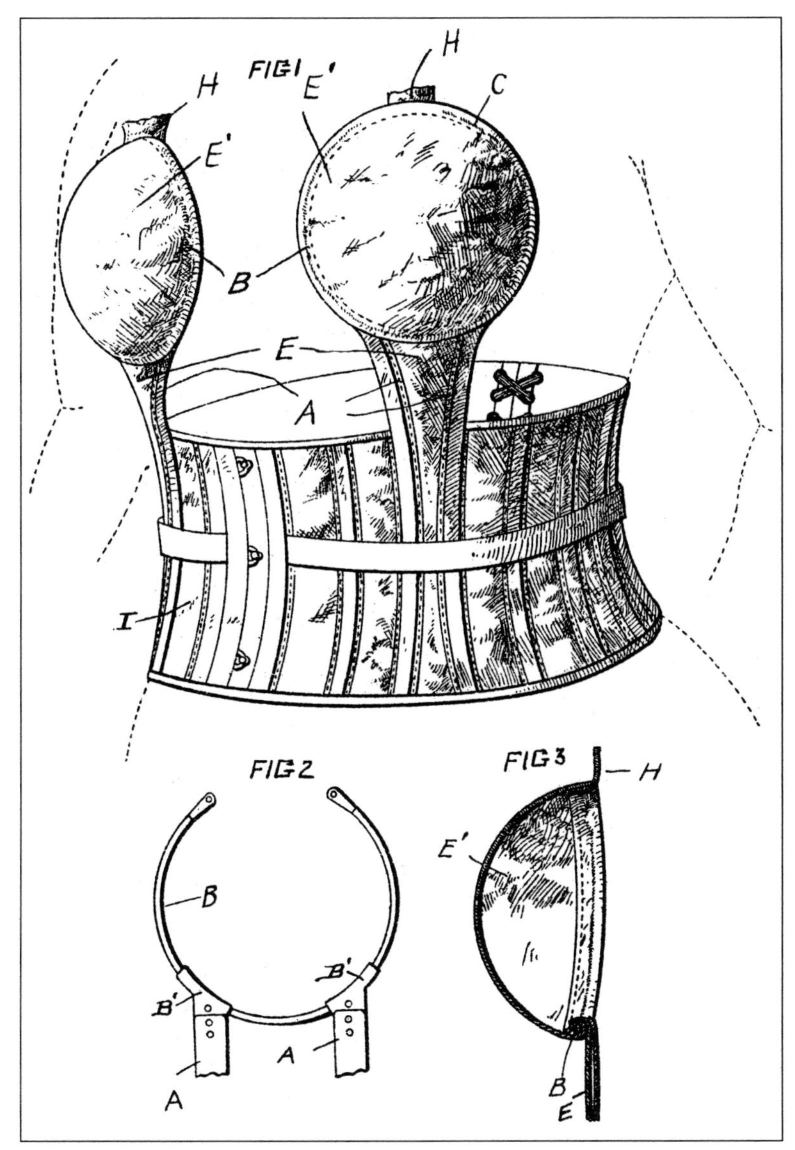

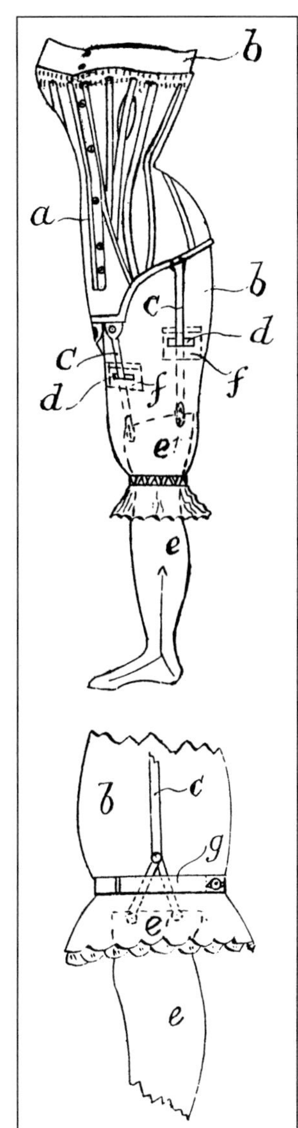

Fulham and her co-designer Herbert Beall, tried to eliminate the perennial inconvenience caused by stocking tops falling beneath the knicker line:[30]

> Suspenders have usually been connected to corsets or belts . . . passing in a downward direction . . . outside the combination, knickers or drawers, their lower extremities passing up under the end of the garment to the top portion of the stocking, to which it is to be fastened . . . To overcome this objection and to ensure the garment remaining in proper position, the combination . . . is provided with one or more slots, or openings, through which the suspender is passed.

Suspenders to be worn over the leg of bloomers and through a slot to attach to the stocking top, designed by a drapery buyer, Emily White and colleagues. Patent.

Above, left: This bust supporter by Lydia Martell and Edward Payne was for breasts of any size. Patent.

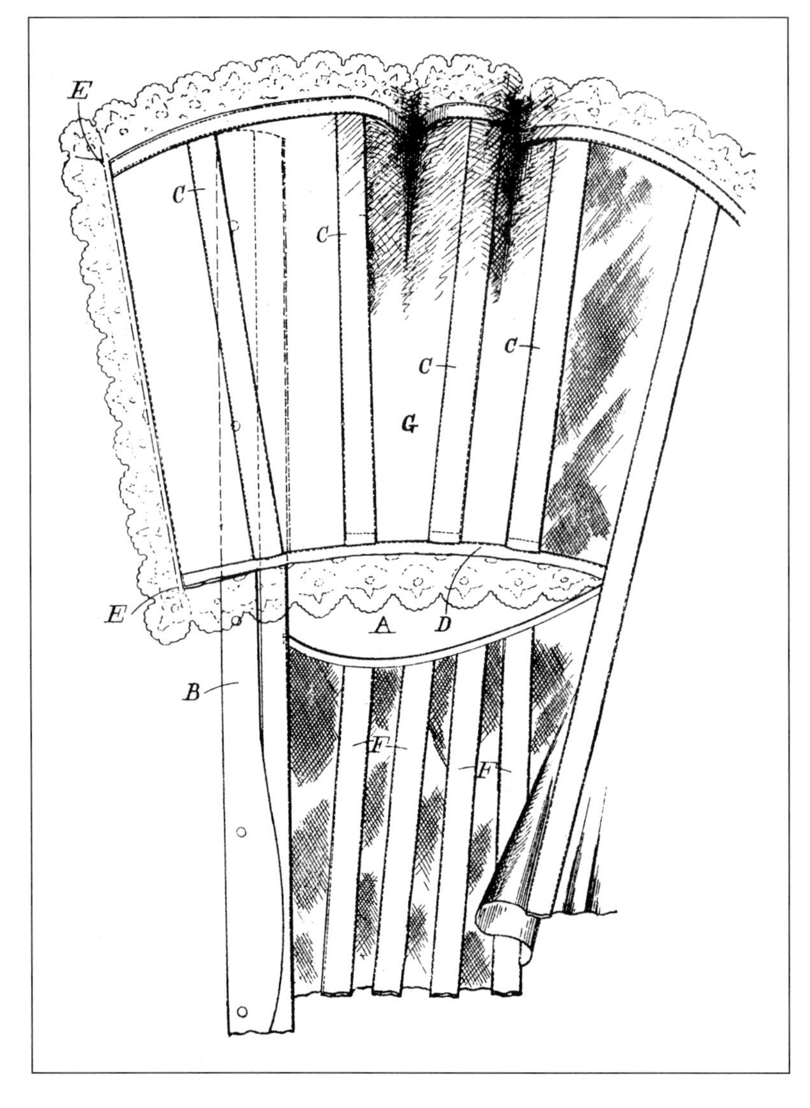

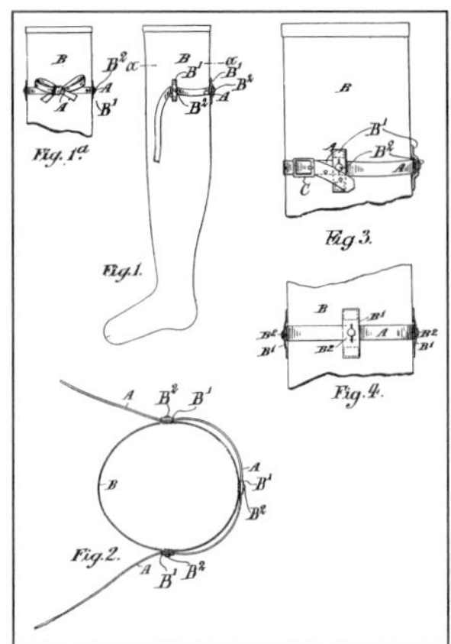

Marian Borsa held stockings up by fastening ribbons around their tops. Patent.

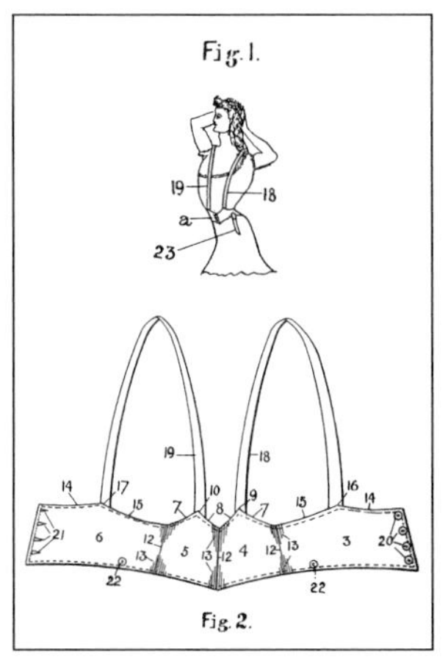

The girdle with shoulder straps by Grace Markell. Patent.

> In order that it can be secured direct to the top of the stocking, without passing up under the lower edge of the garment.

Possibly further anticipating body freedom and the lack of a corset completely, Mrs Marian Borsa from Victoria in the Commonwealth of Australia, devised a method to fasten stockings without any need for the traditional garters or suspenders.[31] She wove into or stitched on to the upper part of the stocking, two ribbons or straps that could be fastened or buckled together. Three years later, in 1908, Grace Markell, temporarily residing at the Hotel Longfellow in Boston, Mass., designed a flexible girdle, rather like a suspender belt but with long shoulder straps, to allow the body freedom to move without 'interfering with the free use and exercise of the body'.[32] It

Above, left: Susan Broadbent's corset to allow freedom of movement for cycling and sport. Patent.

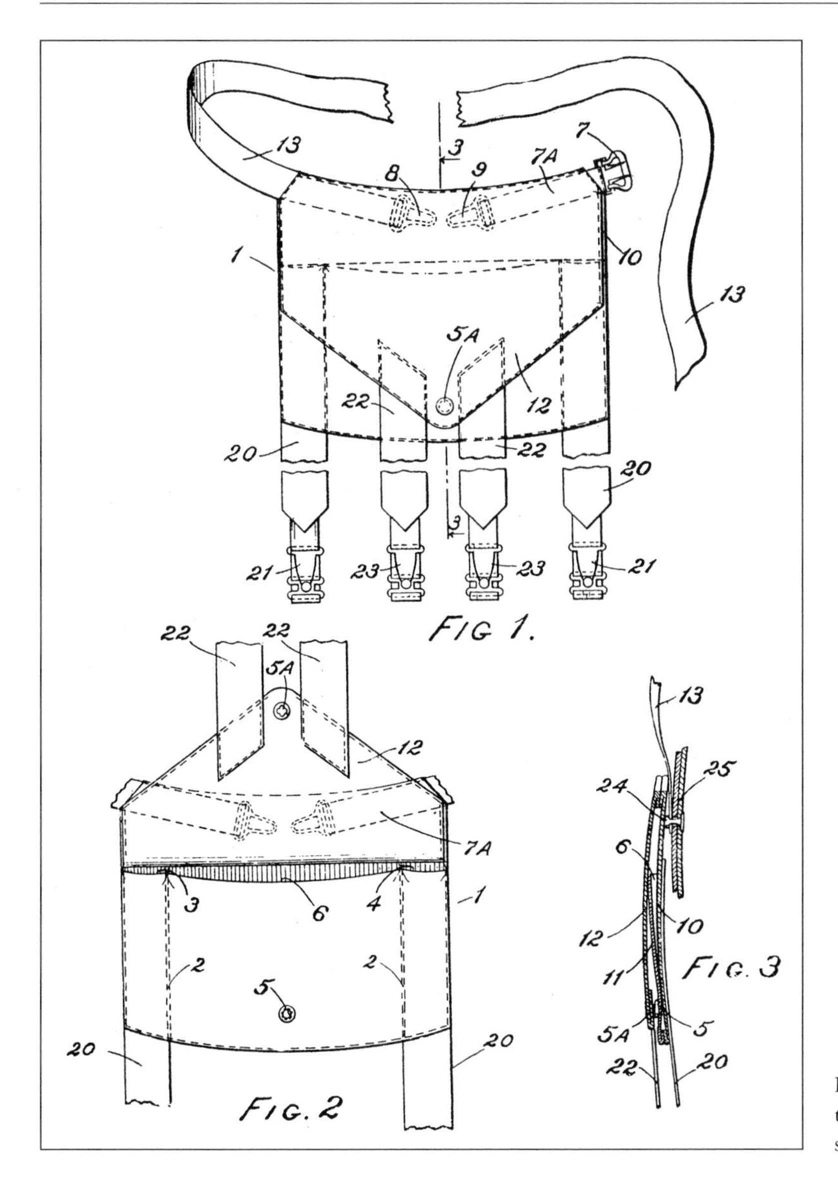

Bertha Ortell hid valuables in the pockets of her garment supporter. Patent.

could be worn for any 'ordinary occupation' but also for 'games and exercise, or by invalids and convalescents'. Susan Broadbent, from Leicester advocated a corset in two parts with a division around the front part of the waist line to allow 'free play to the ribs and intercostal muscles, and to permit full exercise of the lungs and heart which is absolutely necessary for those who take active muscular exercise'.[33] This corset was designed for those with a 'vigorous physical life', presumably the cyclists, tennis players, and those indulging in the other sports now made available to women.

The intimacy of the corset's location was neatly exploited by Bertha Ortell and her ingenious 'Combined Pocket and Garment

Supporter'.[34] Maybe she had been on one of the long Atlantic voyages to America, when she hit upon the idea of combining two flap fastening pockets, one attached to the front the other to the back of a belt worn around the waist. Money, jewels and other valuables could be securely hidden in the pockets, and straps, sewn on to the pockets' corners, acted as suspenders for the stockings.

Sewing

Home dressmaking took on new impetus with the arrival of the domestic sewing machine. Numerous advertisements appeared, like the one in the *Lady's Newspaper* in 1856. Mrs Judd, author of the *Illustrated Handbook for Self-Instruction in Dress-Making*, offered her students 'a course of lessons conveying a thorough knowledge'. The cost was 10*s* 6*d*, with free patterns including a full-size body, jacket and fitting patterns, in French and English styles. Again, women put their practical knowledge of sewing into practice and devised numerous improved arms, needles and methods of stitching. Dress patterns became available and women were able to make clothes on a highly individual basis, which was very popular among those who went out to work such as typists on one of the new typewriters or at the telegraphy exchange. In 1861 Charlotte Craddock[35] devised a method to cut out dresses; hers was a system similar to many of the patterns devised at that period.[36] By measuring the figure in only three dimensions – around the chest and shoulders, the waist and from the armpit to the waist/hips – and using these in conjunction with cardboard templates on which measurements and gradations had been marked, the shape could be worked out. In 1904 Margaret Clarke, a dress-cutting teacher in Glasgow,[37] produced a very detailed cardboard pattern for ladies and children's clothing to enable more accurate shaping of the garment to take place. From Australia, Janet Walker produced a dress stand or tailor's dummy[38] to make the shaping even easier. It had a soft outer layer and was stuffed with kapok and coconut fibre to enable pinning and fitting to suit the shape of the individual.

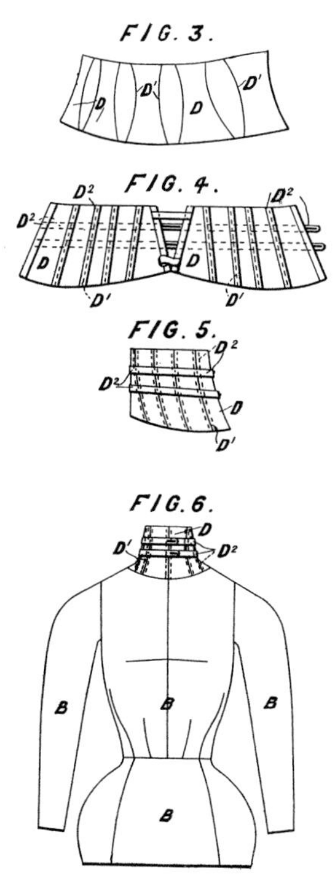

A form of tailor's dummy devised by Janet Walker. Patent.

There were numerous ways to individualise by decoration. Dresses, coats and hats were adorned with flowers, clasps and ribbon ties; skirts were temporarily shortened or lengthened. In 1860, *The Englishwoman's Domestic Magazine*, published by Mrs Beeton's husband Sam, gave many ideas for making and fastening bows for the neck and on skirts. Clasps and fasteners, some ornate, others simple, were devised to shorten and lengthen skirts. Amelia Higgins[39] from Southwark designed flowers and foliage from crêpe and chenille on a silk background for decorative purposes. From New York, Elizabeth Clarke introduced her methods to fold paper into bows to be worn with

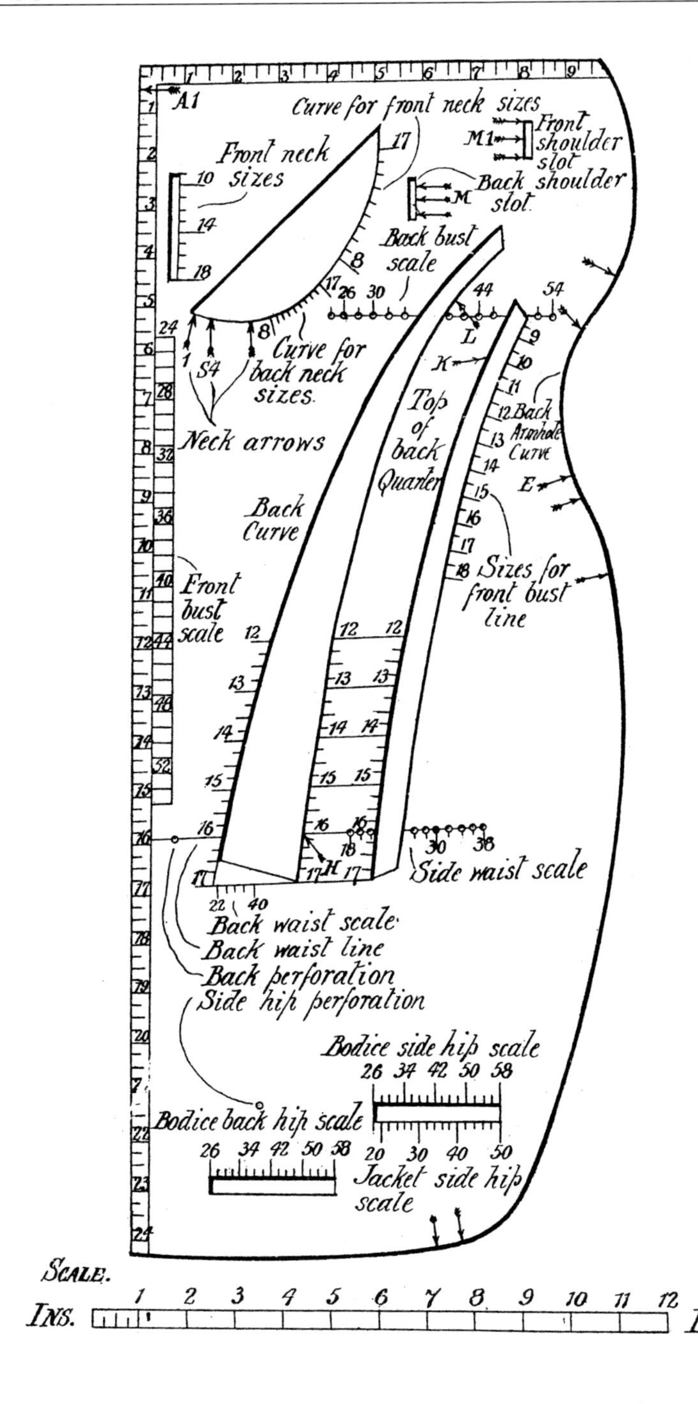

Margaret Clarke's cardboard pattern for home dressmaking. Patent.

collars.[40] French fashion was, even in the mid-nineteenth century, regarded as the epitome of style. Queen Victoria was somewhat in awe of the fashion sense of the Empress Eugenie and tried, often in vain, to emulate her style. Clasps to shorten dresses usually had jaws that could damage the fabric, so Marie Louise Changeur from France introduced her elegant arrangement of cords to loop up the dress in 1865.[41]

Marie Changeur devised a system of cords and loops to shorten a skirt which were preferable to pins and brooches which could damage delicate fabric. Patent.

The tools of the seamstress were not forgotten and there were various inventions for improved pincushions and sewing machine feet. There are many Victorian examples of small storage units to hold thimbles, sewing thread and needles. Many were made out of rare woods to sit proudly on a lady's sewing table and Mary Roy and Louisa Prevett from Penge in Surrey designed one based on an egg and egg cup.[42] A spindle running vertically through the egg cup held the reels of cotton in place and formed a tower with the thimble on top. The tape measure was retractable and coiled around a spring cylinder beneath the spindle, with one end emerging through a slot in the side of the egg cup. The top egg part opened up on hinges to store needles and a cutting blade. Buttons, studs, hooks and faces, all very important to hold a corset together and a dress tightly fitted in the era before the zip fastener, did not escape women's attention and numerous improvements for fastening methods, covering buttons and lacing corsets appeared.

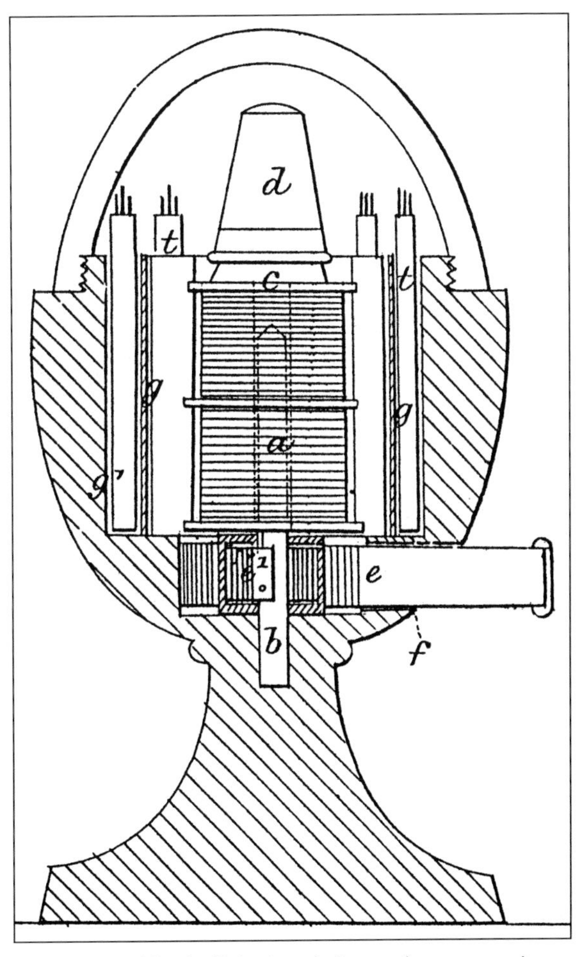

Mary Roy and Louisa Privet's variation on the seamstress's storage unit based on an egg and eggcup. Patent.

Umbrellas

The parasol was part of a smart woman's wardrobe in the 1830s and by the 1880s the umbrella had become a necessary appendage for city life. The variety of design and diverse additional applications of umbrellas make a fascinating study;[43] some were even adapted to become tables, others used for spying. Josephine Julie Besnier de Bligny[44] from France created a parasol/umbrella with a hinged handle to enable the user to alter the angle at which it rested. However, there was an added problem to be resolved, that of how to carry it. In 1867 Elizabeth Brown from Suffolk[45] devised an ornamented, metal plate that was attached to a belt. A chain hung from it and the spike of the umbrella was inserted into a metal ring at the bottom end; the top was held in place by a short chain with a cross bar attached which, when inserted into another metal ring, kept the umbrella in place.

The Bicycle

The influence of the bicycle, from the 1880s onwards, on all aspects of women's lives must not be underestimated. It could be considered as one of the greatest assets to their liberation. Bicycles gave them even greater freedom to travel alone, wherever and whenever they wanted, than that brought by the railways forty years earlier. These cyclists were also part of the new professional class of teachers, telegraphists, typists and nurses, all of whom had to arrive at their destinations in perfect fettle. The long, wide skirt was still regarded as the only appropriate clothing and maintained a woman's femininity, no matter how incompatible with the pedals and wheels of a bicycle. Although Sarah Bernhardt had worn a trouser suit on stage in Paris in 1876, for the masses ankles had to be covered and the wearing of trousers was for them, at this stage, inconceivable. The divided skirt, like the one devised by Florence Jarvis in 1897,[46]

Florence Jarvis's discreet divided skirt for bicycle riding. Patent.

maintained the illusion of a long skirt. Her design 'consists in combining what is known as a divided skirt & knickerbockers in one garment – for cycling or any wear – in place of the knickerbockers, shaped pieces of material are substituted which are fastened to the inside of the skirts at a suitable height & through these the legs of the wearer pass, so forming a knickerbocker garment attached to the divided skirts'. Faced with the problem of a long skirt in wet weather while pursuing sporting activities like golf, tennis or mountaineering, all of which had become popular with the middle classes, Martha Woodall from Chislehurst in Kent devised a method of shortening a hemline. Her 'Improvements in Ladies' Skirts'[47] consisted of a detachable yoke worn at the waist; lining of 'any convenient length' was attached to the inside of it. Press-buttons, hooks, or hooks and eyes were fastened to the outside of the lining and their partners to the inside of the top of the skirt. By lifting up the yoke the skirt could be raised and the fasteners on the skirt secured to those on the yoke's lining.

The need to keep a hat in place, the head dry and the coiffure in shape was fiercely challenged by the growing use of the bicycle and numerous fascinating inventions followed. Even the hatpin, the most obvious way to keep the hat on the head, did not escape re-evaluation. This simple device had its drawbacks: the extended sharp end of the pin could be highly injurious. Annie Smith, a mercantile clerk, and Mary Hamilton Archer, a stock-keeper who worked for George Benson, a sewing machine manufacturer in Cloth Fair in the City of London, devised a 'Hat Pin Protector'[48] in which a small cylinder of wood, metal, ebony or ivory was placed over the end of the pin when it protruded through the hat. Alice Phipps from Northampton had a similar solution,[49] only this time she bent two pieces of wire into a decorative shape in which the point would be retained.

Inevitably, women's inventions devolved beyond the basic hatpin. Elizabeth Reid, Lady Hope of Hyde Park, London,[50] found a solution consisting of a band on a clip spring to fit within the hat and go comfortably around the head. The relative simplicity of this compared favourably with two somewhat cumbersome devices from the continent. The mechanical nature and

Sarah Bernhardt wearing knickerbocker trousers, in *L'Aiglon, a Tragedy*, by Edmond Rostand in 1901. (*Private Collection/Bridgeman Art Library*)

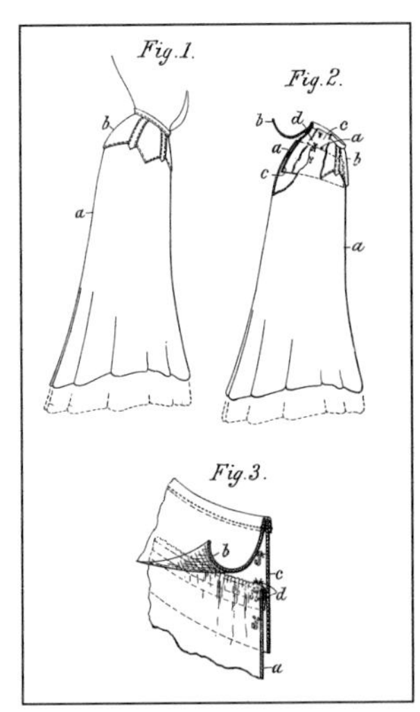

Above: Martha Woodall's method to shorten the skirt from the waist. Patent.
Above, left: Ladies playing golf on Minchinhampton golf course, Gloucestershire, 1890. (*Illustrated London News*)

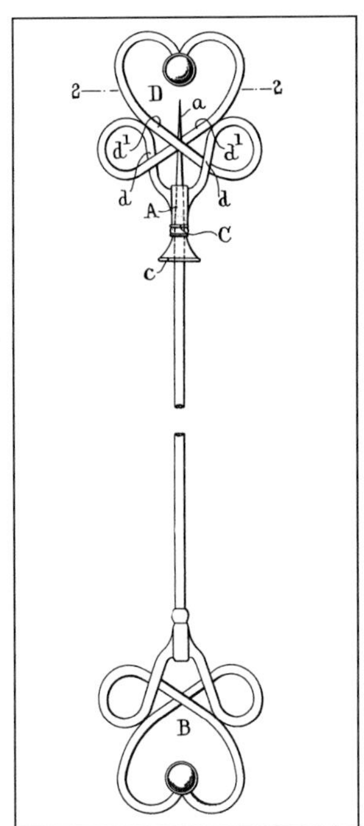

The decorative hatpin and protector designed by Alice Phipps. Patent.
Left: Elizabeth Reid's band and spring clip to secure a hat in place. Patent.

complexity of these creations, whose function had always been simply fulfilled by a hatpin, are possibly a reflection of the high esteem in which the engineer was now held abroad. From Rhöndorf in Germany, Frau Alwine von Otterstedt introduced her 'Improvements in Ladies' Hat Fasteners'.[51] This complicated device, to be hidden within the hat, was clearly only for women with a lot of hair piled on their heads. Two pins were pushed through opposite sides of the inner wall of the hat, knobs were fastened to them on the outside while 'forks adapted to take into the hair' were attached to them on the inside, 'The prongs of one fork being wedged between the prongs of the other fork when both ends meet in the hair, a secure and unobtrusive fastening being thus obtained'. The knobs pulled the forks back as the hat was placed on the head and gently released as it came to rest so that they met within the hair. Marie Tüchler and her three male colleagues in Vienna, Austria evolved a frame, which was placed in the hat and again was reliant on an abundance of hair in order to work: 'two series of . . . curved pins are rotatably mounted in such a wise in a frame to be attached to the hat that by turning them they can be secured in the hair'.[52]

Protecting the hat and hair seems to have been a continental preoccupation. In 1909 Hedwig von Rzewuska from South Tyrol

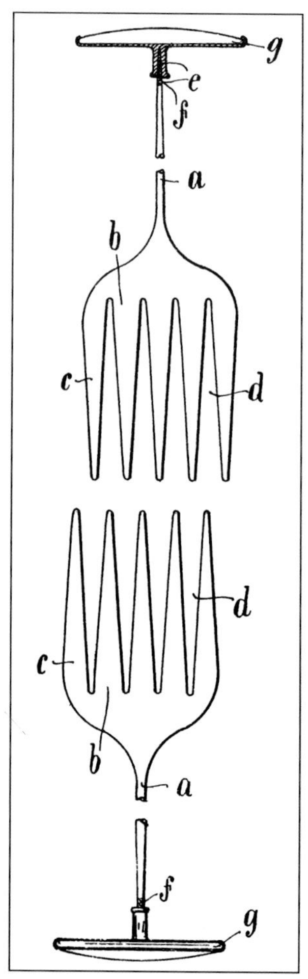

Above: Hair forks devised by Alwine von Otterstedt. Patent. *Left*: The frame to wear inside a hat from Marie and Eduard Tüchler. Patent.

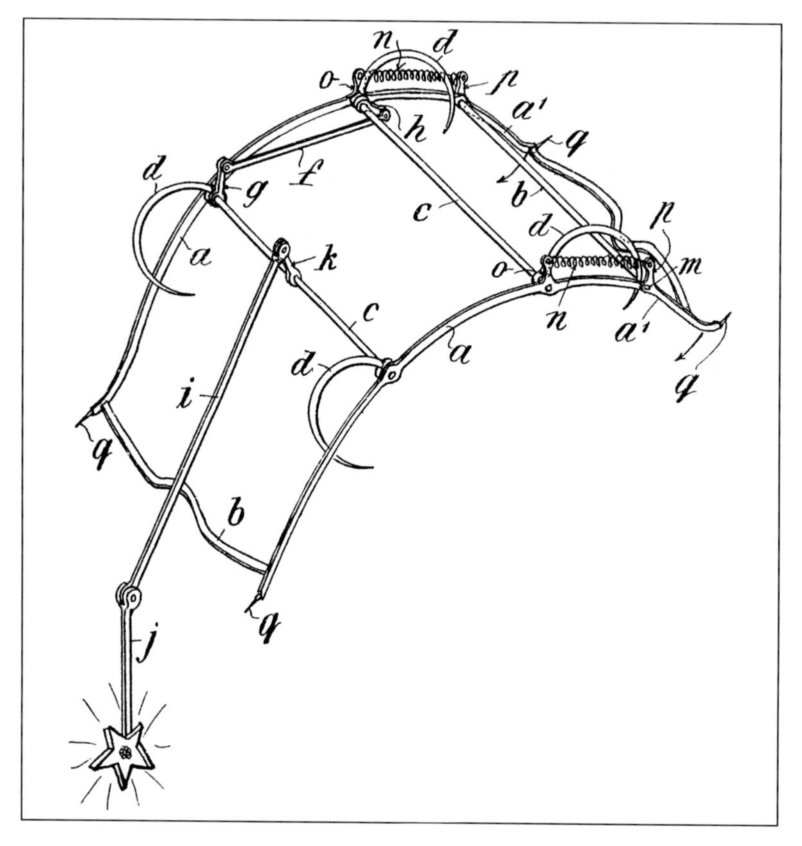

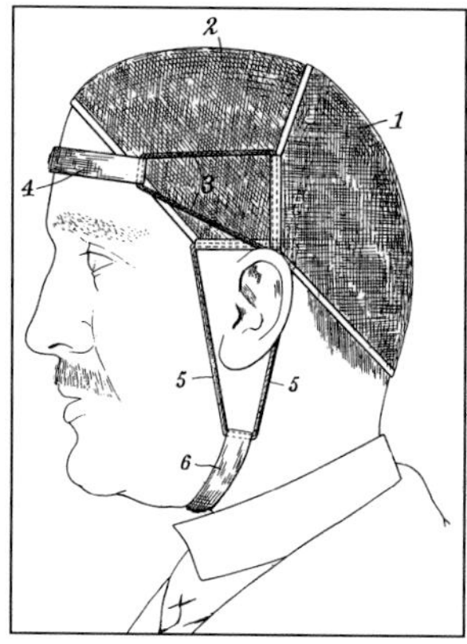

Hedwig von Rzewuska's idea to preserve the coiffeur. Patent.

resolved the problem with her 'Improved Device for Preserving the Coiffure'[53] which, as her diagram shows, was for a man and hardly very elegant. She states that she was adapting the binders for moustaches, which were already in use, for the full head. They were attached to the head with cords and bands: the one placed at the back of the head by a cord passing across the forehead, and likewise one beneath the chin for that at the front of the head. The destruction of a hat by the rain, especially if it was decorated with feathers and elaborate trimmings, could be avoided if Ida Matzen's 'Improved Protector for Ladies' Hats and the like'[54] was used. Frau Matzen from Schleswig-Holstein advocated a type of waterproof bag, which was placed over the hat; hooks around the edges were attached to the brim of the hat.

Ida Matzen's waterproof hat cover. Patent.
Below: A folded umbrella carried over the shoulder with Annie Procter's invention would not be lost. Patent.

Clearly aware of the ease with which things could be misplaced or stolen with an increasingly mobile lifestyle, in 1905 Annie Procter[55] from Manchester devised a method for carrying umbrellas, parasols and walking sticks. 'I provide a bar or strip, or length of metal or other suitable rigid material, having attachments or formations, at or near the extremities thereof, by preference, for the connection of a chain, strap or other flexible band, capable of being readily passed over the arm of the wearer.'

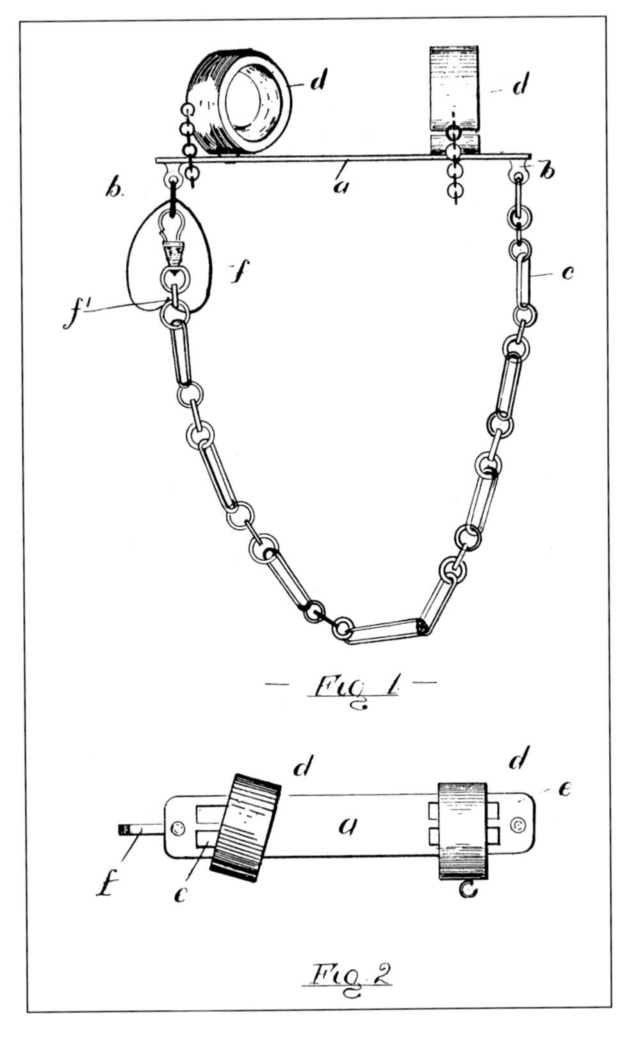

Cosmetics

The names of two women, Helena Rubinstein and Elizabeth Arden, are still synonymous with glamour and aids to beauty. Mr Rimmel from Paris had shown his perfume fountain at the Great Exhibition in 1851 and placed advertisements in the women's papers. Tooth powders to whiten the decaying front teeth were also widely advertised and Maria Farina produced her tooth powder, comprising cream of tartar, calcined alum, orisroot, calcined magnesia and Peruvian bark, in 1856.[56] Clara Lowenberg from New York brought her 'Composition for Beautifying the Complexion' to Britain in 1866.[57] Like many potions and lotions, before and since, she claimed that hers would have 'for its object the beautifying of the complexion, and is distinguished for its extremely bland, purifying, harmless and soothing effects upon the skin and by acting upon the pores and minute secretory vessels; it expels all impurities from the surface and effectually all redness, pimples, spots, blotches and such like imperfections of the skin'. Her potion contained pulverised egg shells, calcined magnesia, terra alba, marsh mellon root, orisroot, gum benzoin, rose water, otto of rose and cascarella root.

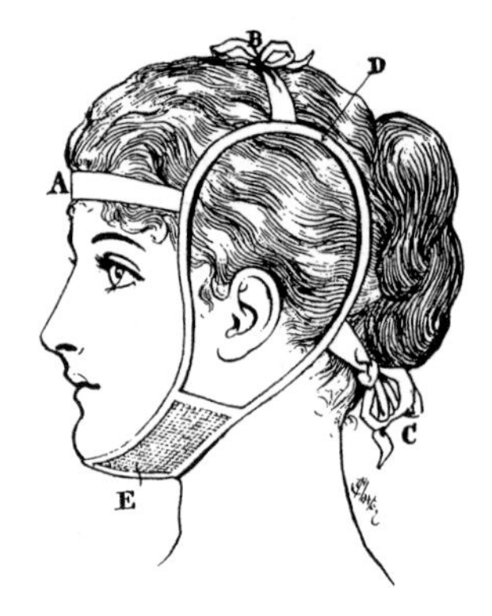

A double chin might be cured if Adelaide Turner's device was worn. Patent.

It was, though, Helena Rubinstein and Elizabeth Arden who invented the concept of the modern beauty parlour and its ensuing products, both evolving into huge multinational companies. Helena Rubinstein was born in Poland, studied medicine for a short time in Switzerland and was interested in dermatology, before going to Australia in the 1890s. She opened a shop in Melbourne to cater for the increasing number of Australian women for whom there was little access to skin preparations. In 1908 she came to London and opened a salon, followed by another in Paris in 1912 and in New York in 1914. At the same time Canadian nurse, Florence Nightingale Graham, was drawn to the metropolitan excitement of New York in 1908 where she formed a partnership and opened a beauty salon called 'Elizabeth Arden', on 5th Avenue. Soon afterwards the partnership split up but Florence Graham decided to keep trading as 'Elizabeth Arden'. She and her team devised and marketed, in the beauty parlours, cosmetics desirable to women eager to glamorise themselves. Helena Rubinstein and 'Elizabeth Arden', while pursuing similar markets, remained rivals throughout their lives. The use of modern cosmetics at this time often met with disapproval from the status quo and the women who patronised the salons tended to be those from wealthy backgrounds but with a rebellious streak.

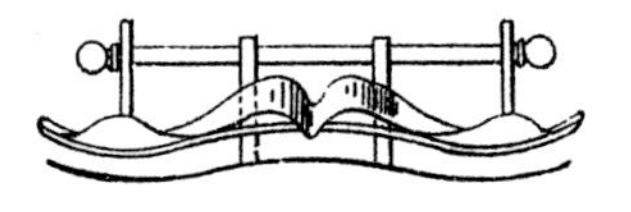

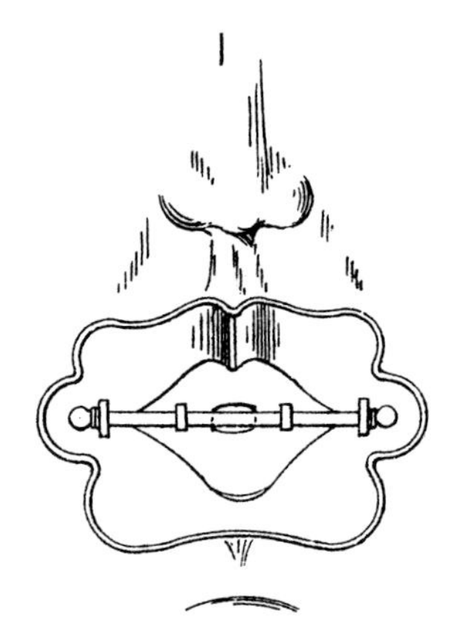

Anna Leroux advocated wearing plates to alter the shape of the mouth. Patent.

A more invasive approach to achieving beauty was represented by the device for 'Curing a Double Chin' designed by Adelaide Turner in 1904.[58] Adelaide Turner, a 58-year-old artist and wallpaper designer, ran her own business in Chiswick, West London and unusually employed her husband as her manager. Her patent claims that her device, made of non-elastic, ventilating material on a frame would be a 'new or improved double-chin cure consisting of a frame of pliant material . . . attached to said frame in such a position so that when tied to the head, said material will press under the chin of the wearer, substantially as described'. In Paris a widow with a similar aim, Anna Leroux, introduced a 'Modifying Apparatus for the Mouth'[59] in which a plate was placed inside the mouth, behind the lips, and fixed in place by a bar across the front. 'The body of the apparatus, viewed from the front, has two uprights or racks, each provided with three holes in order to allow the apparatus to be tightened as desired by sliding the bar to the degree required.' A tooth at the upper edge was very prominent in order to accentuate the curves of the lip and to 'give to the contour of the worst shaped mouth, the harmonious curves of a well modelled mouth'. There was a hole in the plate for breathing and a rubber tube could be inserted for feeding purposes.

The Personal

While Amelia Bloomer gave her name to bloomers, from which modern knickers developed, other female concerns also received attention. There are many patent applications for sanitary towels, often cumbersome and appearing extremely uncomfortable. Naturally, women did devise methods to deal with the monthly cycle. Elizabeth Cumming designed a simple 'Improved Sanitary Towel'[60] which comprised an antiseptic pad attached at its four corners to a belt. Alice Lang from Birmingham[61] devised an 'Occasional Sanitary Shield' which received approval from the medical profession; it was stain-proof and worn with a sanitary towel to prevent soiling the undergarments. It was made of 'jaconet', a light, cotton fabric originally from India, which was perforated at the top to prevent the body getting too hot and weighted at the bottom with waterproof sheeting. It was fastened around the waist with a band and buckle. She refers to other methods using washable material on to which waterproof fabric was buttoned, like an apron, and claimed her invention was different from these. Annie Hebblethwaite from Leeds[62] designed knee protectors for knickerbockers, which were fastened around the knee beneath the knickerbocker to prevent them wearing out.

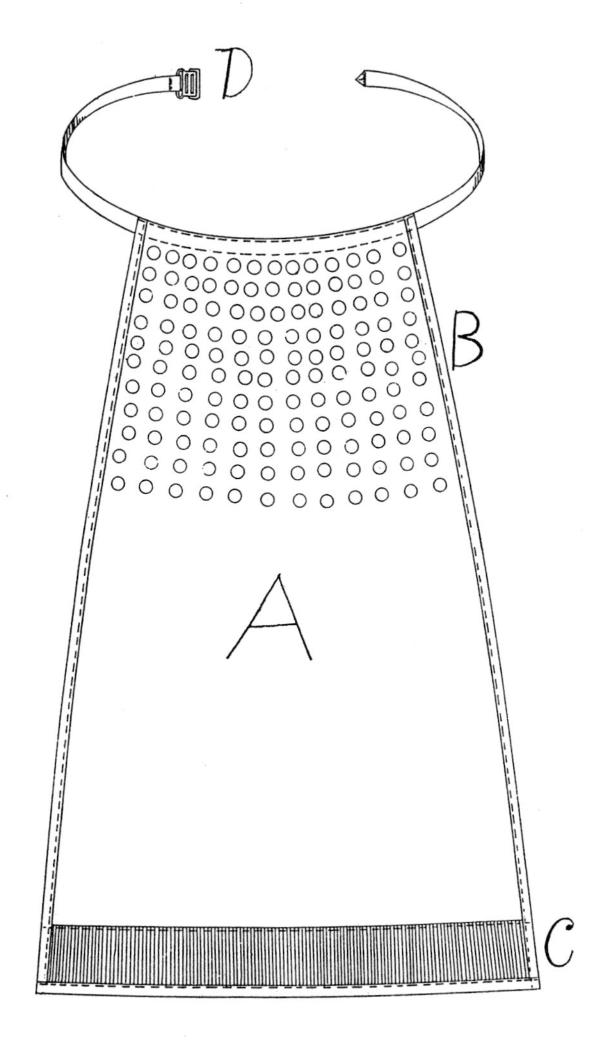

The stain-proof sanitary shield by Alice Lang. Patent.

CHAPTER FOUR

Domestic Challenges and Home Improvements

The industrialisation of the nineteenth century brought enormous and profound economic and social changes as people moved from the countryside to the towns. In 1801, 22 per cent of the British people lived in towns; by 1851 the proportion had increased to 50 per cent; and finally, to 77 per cent at the century's end. Manchester's population expanded by more than eight times during the century.[1] This rapid growth of the industrial towns and cities, the enormous technological changes and the reassessment of religious belief combined to create unprecedented upheaval in the structure of society. A new middle class, made wealthy from the many new industrial ventures, emerged, while millions of working-class factory and domestic workers lived in poverty. They worked long hours and, before legislation and the trade union movement, in appalling conditions. The massive increase in homes for the new middle classes; the conditions endured by the poor; the lack of understanding about the spread of disease from poor sanitation and hygiene became great focuses of attention and problem solving for the Victorian inventor and philanthropist. Women too were part of this inventor's paradise.

In 1851 thousands of people travelled to London to the Great Exhibition in Hyde Park and saw, for the first time, all variety of gadgets, machines, furniture, dinner services and much more from Britain and around the world: this was the first international exhibition of manufactured goods. The archetype of the happy family and cosy home was perpetuated by photographs of Queen Victoria and Prince Albert with their growing family. Theirs was the first royal family whose photographic images were available to the press and their ideal became something to be emulated. The influence of the Gothic-style Balmoral and Italianate Osborne royal holiday homes, is apparent in the thousands of houses built with pointed arched windows, gabled or flat roofs, balconies and pillared porticoes.

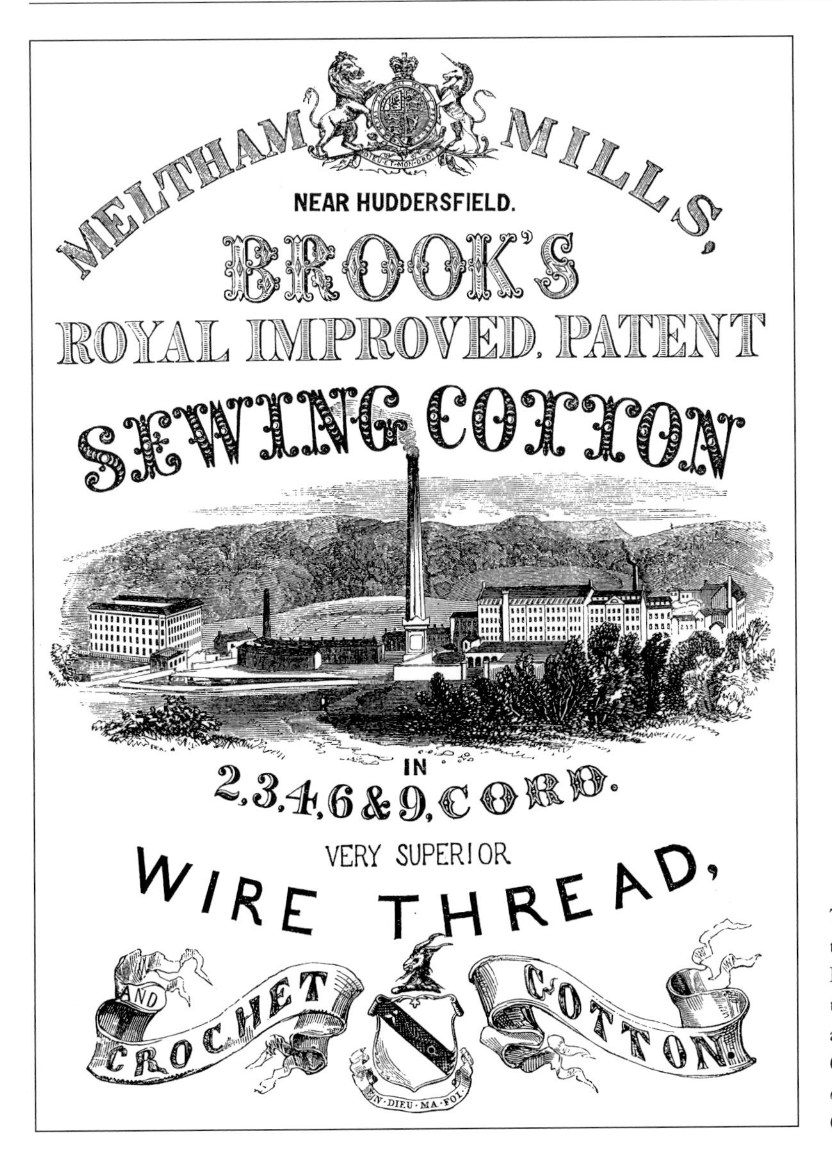

Thousands of women worked in the cotton and woollen mills of Lancashire and Yorkshire during the Industrial Revolution. This advertisement is from the *Official and Descriptive Catalogue of the Great Exhibition of 1851*. (*Author's Collection*)

The Victorians loved gadgetry, especially for domestic purposes. In most middle-class, urban homes the talk of invention must have been an almost daily occurrence. Magazines, like *The English Woman's Domestic Magazine*, launched by Sam Beeton in 1852, and *The Table*, founded by Agnes Marshall in 1886, doled out advice, advertisements and articles to the housewife with time and the ability to read. There were manuals as well, most notably Mrs Isabella Beeton's *Book of Household Management* (1859). Recognising the plight of the lonely woman inexperienced at running a home, but eager to impress her husband and his friends, Mrs Beeton administered advice on the hiring of staff, looking after children, planning rooms, arranging furniture, keeping clean, and dealing with legal and medical matters as well as recipes for breakfast, lunch and dinner for adults, children and invalids. From the 1880s, people like the entrepreneurial Mrs Marshall, who opened Marshall's School of Cookery in London in 1885, offered cookery courses.[2] Attitudes to food changed too as more varieties were shipped, preserved in one of the new refrigerators or vacuum canning processes, from all over the Empire and the USA to appear in shops like The Home and Colonial Stores.

In Britain, where domestic labour was cheap, the mistress of the house usually remained above stairs to be consulted by her cook and housekeeper; hers was much more the role of manager, whereas in Germany she tended to work alongside her domestic staff, and in the United States the pioneer wife was being supplanted by the city dweller. Following the sudden upsurge in interest in the countries of the Empire, instigated by Disraeli in the 1870s, working-class families were encouraged to emigrate to a new life of fresh air, space and opportunity. In due course, a number of inventions appeared from these women. The British housewife's upstairs role did not deter her from inventing all variety of household gadgets. By the 1880s there was an increase in patent applications in Britain by German and American women, whose approach, reflecting the later industrialisation of their countries, tended to involve more engineering and factory-produced items than their counterparts in Britain. This trend was also reflected in their more robust style of technical drawings and diagrams.

Women invented new types of chimney stacks and methods of cleaning them, and they improved polishes, window blinds and curtain hangings, stoves, pans and utensils, candles and gas lights, dress stands, furniture, fire escapes, machines to wash clothes and dishes, and much more. They made improvements to the nursery and put their hand to complete interior decoration schemes. Ideas for the recycling of sewage waste were not forgotten in the aftermath of Edwin Chadwick's *Enquiry into the Sanitary Conditions of the Labouring Population of Great Britain*. Some inventions were simple and uncomplicated but often there was a confusion of purpose and design, which dogged the Victorian approach to design. Admiration of the multipurpose gadget could overwhelm the need to fulfil the fundamental function as various uses, some rather unlikely, were claimed for one article. Christina Hardyment observes that the Victorian enthusiasm for invention and mechanisation could get in the way of the actual activity, making it actually more complex; in reality many tasks were easier left to be done by hand. This trend continues today with the endless so-called improvements on offer at household exhibitions with the aim of doing, for example, what a basic mop and bucket or duster can do. However implausible these machines and gadgets might seem, they are fascinating and 'do illustrate the fertility of the inventors' minds, and the receptive attitude to the idea of labour saving in the home.'[3] Every aspect of urban living was scrutinised by women.

Lighting

Poorly lit streets and buildings, still regarded as a problem today, concerned Elizabeth Perryman from Soho, London in 1809 when she devised her 'Improved Street and Hall Lamp, and the necessary apparatus for expediting the trimming, lighting and cleansing the same'.[4] Although the combustible nature of gas had been recognised by the Revd Clayton in Yorkshire as early as 1684, the progress to wide-scale introduction for street lighting and, later, in the home and factories was slow. It was used for public lighting first. In 1814 the Gas, Light and Coke Co. had laid pipes in Westminster and by 1849 there were gas-works in most towns.[5] Elizabeth Perryman's invention, 'the result of careful and practical experiment', was a gas-powered lamp to increase the amount of light produced without increasing the quantity of gas burnt. The upper part of the lamp was to be made of a curved reflector 'which receiving all those rays of light which in the common lamp are absorbed . . . by a well known law of catoptrics, at the same angle at which they are received . . . distributes this . . . additional quantity of light'. The lower part of the light was to be made of glass 'of such shape as that all its parts shall stand respectively as near as may be at right angles . . . to those parts of the reflector from which the rays proceed are to pass through it'. In addition to these benefits her lamp could be lit without a ladder, thus saving risk and time.

Even though gas was widely available by the mid-nineteenth century, many mistrusted it. It gave off an unpleasant smell and

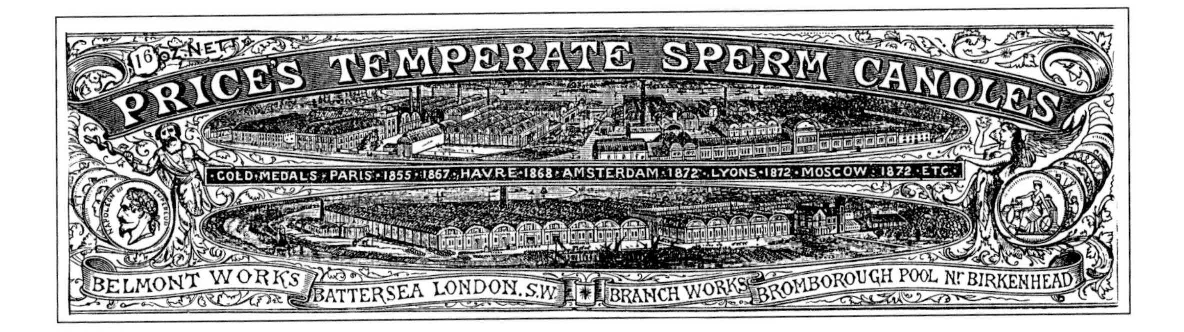

A trademark design for Price's candles, 1876.

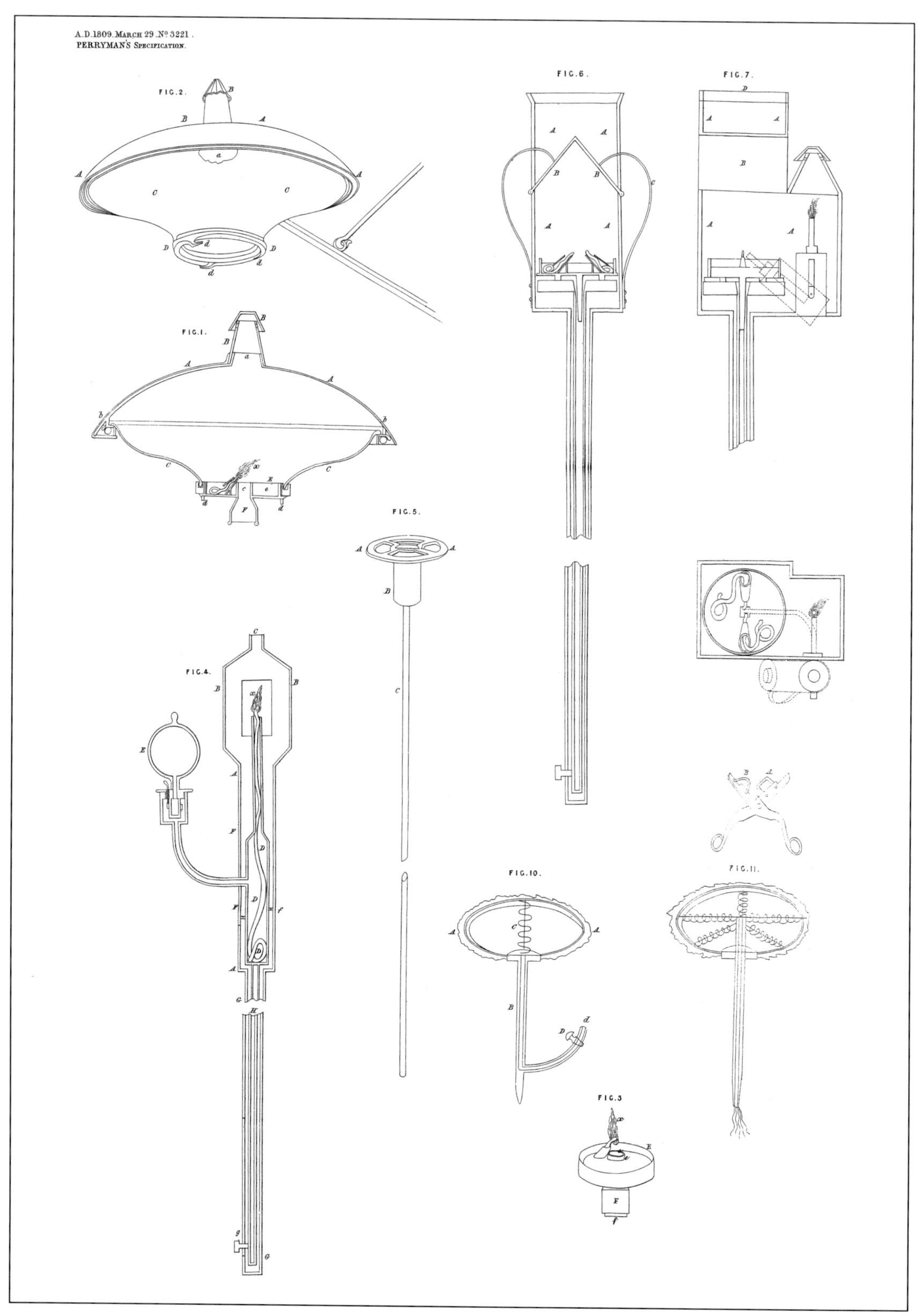

Elizabeth Perryman's street and wall lamp. Patent.

flames from the lamps could be dangerous if left unguarded. Candles were still often preferred and companies like Price's produced all varieties, including a type that would not spill when moved and others designed for a warmer climate. There were many devices for holding candles or trimming them, and there were candle lamps like the one by Marie Louise Lise Beaudeloux from France.[6] Throughout the century the influence of French design in fashion, food and the home cannot be over-estimated; Queen Victoria herself was enthralled by the French fashions. Marie Beaudeloux's device resembles that of a modern lipstick, in that the candle, no matter how far it burned down, was continually pushed upwards by a mechanism at the base of the coloured tube in which it stood, so that the flame was permanently just above its tip. In 1859 Mrs Elizabeth Steane from Brixton designed a protector to stop candles dripping, no matter at what angle they were held.[7]

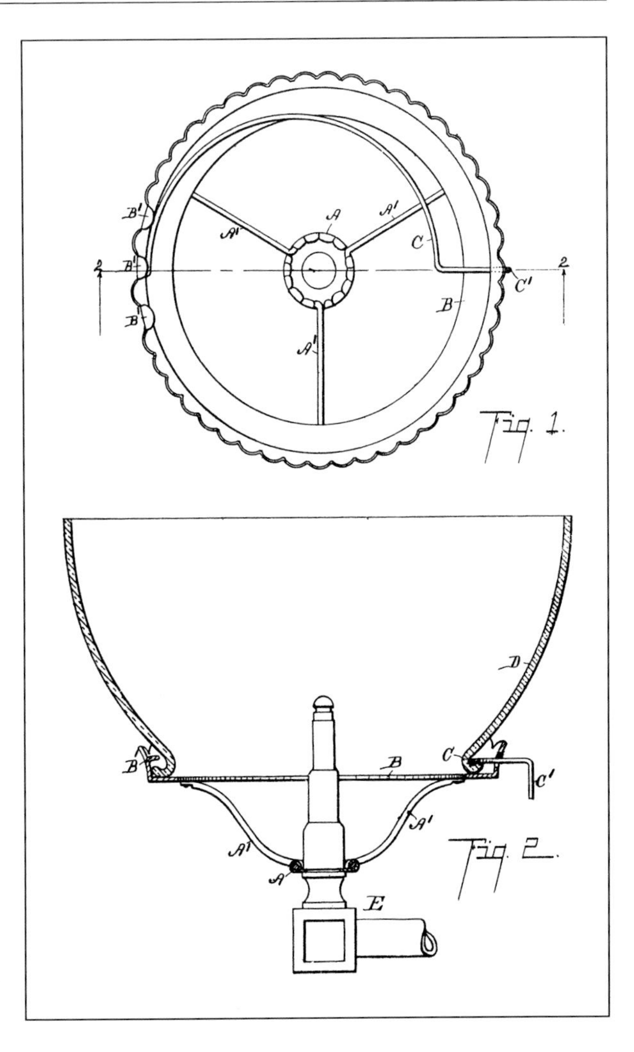

Almost one hundred years after Elizabeth Perryman's street lamp, Bertha Miller Windoes from Michigan registered her invention for improved globe holders for gas burners,[8] which could also be adapted for use with an electric light bulb. Thomas Edison had patented his incandescent electric lamps, in New Jersey in 1879 but electricity, like gas in its day, was slow to catch on, as gas lighting was preferred in many homes. Although she had been shown the merits of the electric light bulb, even the Queen preferred the cosy warm light emanating from a gas burner to the comparatively harsh qualities of that produced by electricity. Clearly influenced by the Art Nouveau movement, Josephine Marie-Louise Fleming, living in a château in northern France, produced a beautiful lamp[9] which had a wick and could be fuelled by petrol, oil, alcohol or paraffin. Her specific method of assembling the parts meant that the fuel would not block the internal pipe work. The technology apart, the style of this illuminated lamp would have enhanced any salon.

Globe holders for gas burners or electric light bulbs devised by Bertha Windoes. Patent.

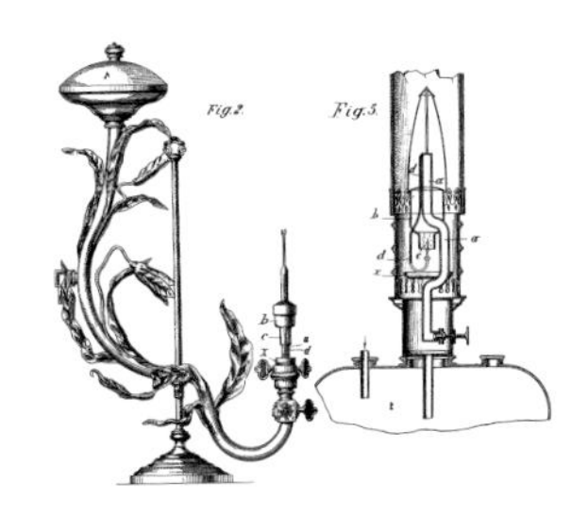

Josephine Fleming's elegant Art Nouveau-influenced table lamp. Patent.

Heating

The image of the Victorian industrial town, smoke billowing from the chimneys of its factories, houses and steam trains, enveloped in thick dirty smog, is familiar. Coal, essential to power and heat the industrial revolution, brought with it a host of problems. Until 1842 women and children worked under ground in the coalmines of Britain, in what must have been some of the most dangerous and dirty work of the period. The Mines Act 1842 ended the practice of women and children under ten years old from going beneath ground; various Factory Acts from 1819 onwards also regulated the ages of children employed and the hours and conditions in which they worked. In the home, however, housemaids got up before dawn to heave buckets of coal upstairs to fireplaces in every room of their mistresses' homes. Working-class women had to make sure there was enough coal for the fire on which to cook and keep warm.

Coal was dirty to handle, chimneys got blocked with soot and could catch fire, and windows and furnishings got dirty from its ash and dust. Between 1853 and 1856 various Smoke Abatement Acts were introduced. It is not surprising, therefore, that women became involved in devising ways of coping with coal in the domestic environment. At the same time as Elizabeth Perryman was working on her street lamp and Rebecca Ching was perfecting her worm-destroying lozenges, Elizabeth Bell, a spinster from Hampstead and later Blackheath, was granted three patents to introduce an 'artificial method of sweeping chimnies and to construct them in such a way as to 'lessen the danger and inconvenience from fire and smoke'.[10] In 1803 she proposed two methods of constructing and cleaning them, which she refined four years later. Furthermore, she clearly claimed ownership of the idea in her patent: 'Certain improvements in an artificial method of sweeping chimnies invented by me'. Miss Bell designed two adjustable, semi-circular iron frames that could be adapted to suit the variety of shapes of the 'new' chimney pots. The chimney could be thus swept from the bottom and, by pulling a rope or chain hoisted over a set of rollers on the top of the frame, and hanging down the chimney, cleaning could take place very quickly. Lever brushes could be added where appropriate and a fire extinguisher attached to the chain. In 1880 Clara Rouse from Kent devised a 'Soot Screen for use in Sweeping Chimneys'.[11] Matilda Stewart Barron exhibited her noiseless coal scuttle in the Women's Pavilion at the Chicago Exhibition in 1893 to dampen the noise emitted when transferring coal from the scuttle to the fireplace.[12] Finally, in 1905 Pauline Grayson, a London artist,

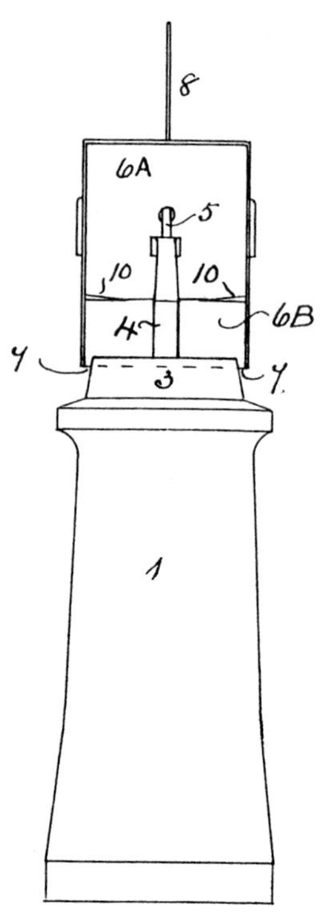

Chimney cowls from Martha Helliwell. Patent.

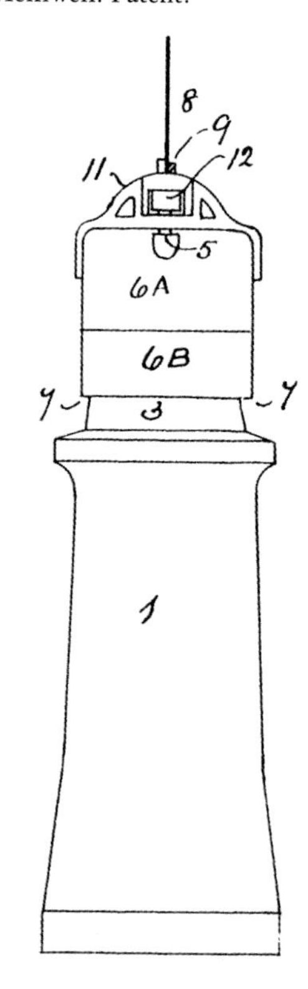

achieved a near-smokeless fuel by mixing fine coal dust with plaster of Paris or cement to form a block. Fifty-three-year-old Martha Helliwell[13] was married to a stonemason, which probably gave her the versatile knowledge of chimneys, brick and stone, for her 'Improvements in and Relating to Cowls for Chimney and Ventilating Shafts'. In her invention the cowls were designed to rotate or incline and be self-regulating dependent upon the direction of the wind.

Scrubb's advertisement for soap, 1905. (*Illustrated London News*)

Another novel aspect of the Victorian middle-class home was the prospect of hot running water for the bath, laundry and kitchen. The geyser, so called after the hot water springs in Iceland, was invented in 1868 by Benjamin Maugham. In it, water was piped in a coil through a cylinder heated by gas.[14] This led the way to a variety of geysers being invented, like the one by Florence Gill[15] which provided a more efficient means of heating water for the bathroom. In her design, the space above the burner was fitted with asbestos or firebrick to retain the heat.

The contrast between middle- and upper-class toileting arrangements and those of the working classes was stark. The raw image of the tin bath placed in front of the range, for the weekly family bath, has become a symbol of the poverty and hardship endured by millions. Laura and John Edwards[16] from Pontypridd, in the heart of the South Wales coalfield, aiming to solve at least the problem of the lack of modesty for miners while bathing, devised a screen, attached to the mantelpiece, which extended around the bath. 'The particular object of this invention is to give a collier privacy to wash his body with both hands together when standing nude in his bath tub by the hearth, at the same time screening him from draughts: it being common knowledge that it is very rare to find a bathroom in a miner's house the usual practise being to wash in a tub on the hearth openly.'

Water could be heated in Florence Gill's geyser. Patent.

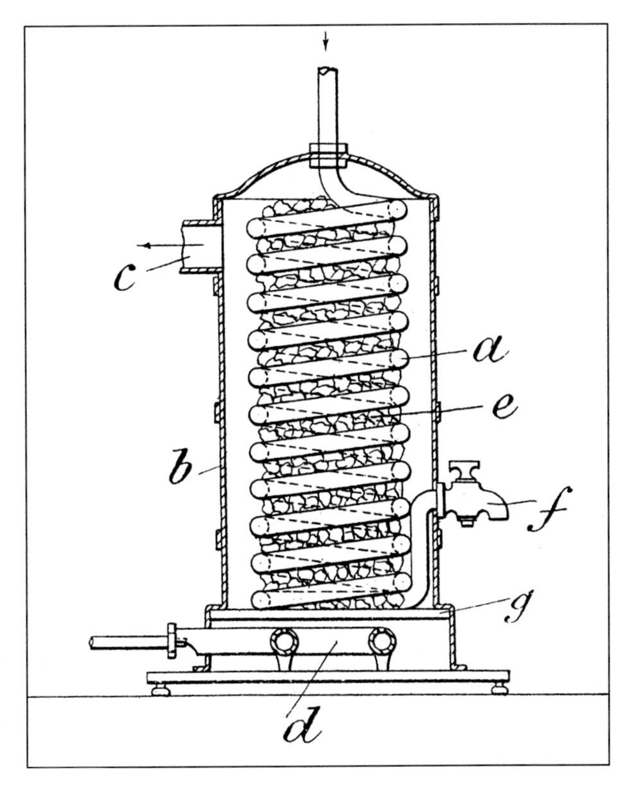

Sanitation

One of the greatest threats to health in the nineteenth century, emanating from the proximity of urban living conditions, was that of the lack of sanitation. The relationship between poor hygiene and the spread of disease was not understood and so cholera, typhoid, smallpox, scarlet fever and tuberculosis were endemic and classless. In 1842 Edwin Chadwick reported to the government on the results of his *Enquiry into the Sanitary Conditions of the Labouring Population of Great Britain*, which was instrumental in advancing the understanding of public health. Public washhouses were opened with baths and laundry facilities for women to wash and dry their families' clothes: a Victorian launderette. Here, too, families could bathe and use the swimming pool. Huge sewerage schemes began. One of the first sewers to be laid was beneath Fleet Street in the City of London. Many people, however, including Prince Albert, still believed that human waste could be used beneficially. When Osborne House on the Isle of Wight was being built he designed a scheme to pump the waste from the house into the sea and back on to the farmland as water and fertiliser. Mary Beale from Barnsbury in London and John Beale from Maidstone were also oblivious to any health risk from using human waste in their instructions to make manure.[17] In an era before the water closet was widely available and the chamber pot a convenient night-time receptacle, they took 60 parts of dry night soil and mixed it with 30 parts of lime, and ½ of soot, 2 of common salt and ½ part of dregs of asafoetida. When suitably mixed into a powder it became not only a fertiliser but a weedkiller too and was sprinkled on the land: 'inducing a rapid growth of the crops it destroys all animal and insect life so detrimental to growing crops'.

In spite of the improvements, a consequence of the sewers was the unpleasant odour which emanated from them and permeated nearby houses. A San Francisco physician, Elizabeth Corbett,[18] hoped to combat this in 1878 with her gas street lamp in which the

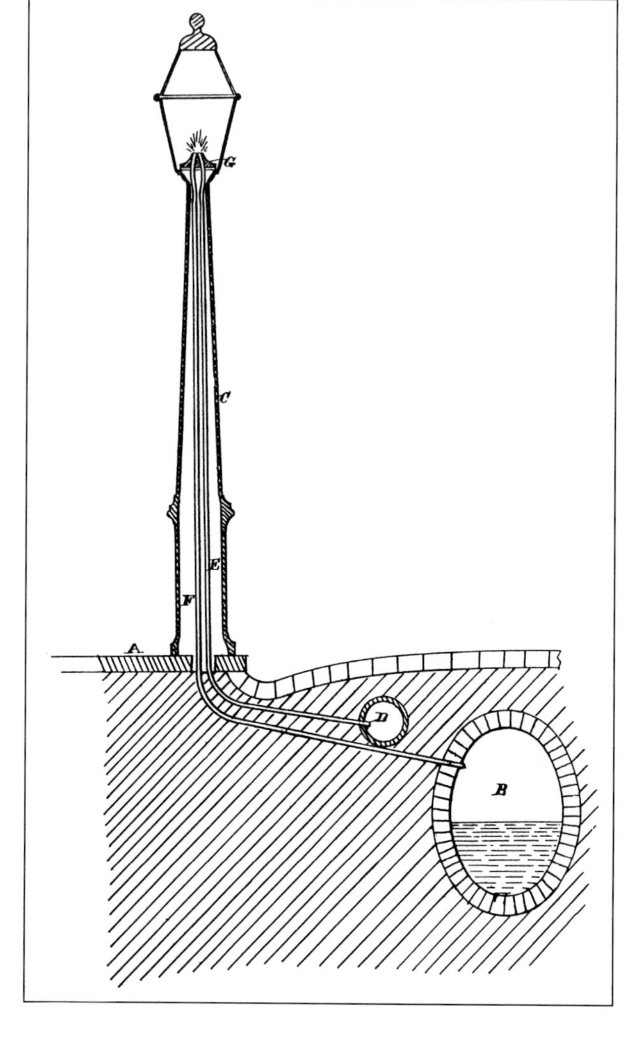

Elizabeth Corbett designed a street lamp that would burn off fumes from the sewers. Patent.

noxious fumes from the sewers were burned off. By running a pipe from the sewer to a row of gaslights, the fumes would be 'conducted to (the lamp) and delivered into the flames and thereby destroyed'.

Earth and water closets were being installed in new homes by the 1880s, but were still regarded as a luxury; commodes and buckets and then the outside lavatory remained the norm until the First World War. Better knowledge of hygiene still did not resolve the dilemma of what to do with the slop bucket. Jessie Deans from south London designed a night stool/commode to address these issues.[19] Instead of a ceramic bowl beneath a wooden seat, which would harbour germs, hers was like a tall, earthenware bucket on which to sit. The top of it was rolled over to form a seat with a lip through which the contents could be poured out. There was a small ridge around the inside of the rim on which a lid could rest. In previous commodes the slop bucket would have to be removed and was rather a disgusting procedure: smelly and unsightly; with this method the whole lidded bucket could be carried away and emptied elsewhere. Also, it was easy to keep clean, as there were no sharp edges.

Slops could be hygienically disposed of in Jessie Dean's bucket. Patent.

The Laundry

Doing the laundry has traditionally been women's work: gathered in groups by a stream or river, at the village pump, collecting dirty linen from a wealthy mistress to wash it in one of the new municipal laundries, they have, for centuries, rubbed cloth against rocks, stones and wooden washboards. Immersion in cold water, constant bending and carrying heavy loads have taken their toll, over the years, on women's hands and backs. So it is unsurprising that they have searched for new methods, gadgets and mechanisms to lessen their burden. Mary Buchanan[20] from Glasgow made protective gloves out of two pieces of India rubber. They were cut into the shape of the hand, stitched down the sides and coated with the India rubber to make them watertight. Numerous soaps and cleaning products, like that by Maria Rowland from Acton, appeared.[21] Her improved soaps and

On Washerwomen, nineteenth-century lithograph, English school. (*Private Collection/ Bridgeman Art Library*)

detergents were for personal and laundry use and consisted of a solution of soap dissolved in hot water into which a combination of ammonia, liquid hydrocarbon, turpentine, coal-tar, naphtha, camphine and benzene was dissolved. Essential oils and scent were added to eradicate any chemical odours when used for toilet soap. On a more industrial basis, Agnes and John Wallace from Renfrewshire[22] came up with a new method of bleaching, washing or cleansing fabrics using hot air and gas pumped through a dash wheel into a stuffing box containing the textiles and bleach, which improved and speeded up the process. And in 1859 Elizabeth Merrell,[23] a tin and zinc plate worker from London, came to devise a machine to wash clothes using electricity. 'I form . . . or furnish washing or cleansing apparatus with conveniently placed and suitably connected elements, and on these being excited by the washing or cleansing liquor, water, or suds employed, electric or galvanic action will be developed, and will act in turn upon the suds, water or purifying liquid.' Through a complex series of procedures the elements were placed in a zinc case with corrugated sides, and then into the washing vat containing soapy water and the dirty clothes. The electric current caused the corrugated sides to move and agitate the soapy water thus cleaning the clothes. Mary Brown, a teacher from Brighton,[24] exhibited her washing copper at the Chicago World Fair in 1893. In it a propeller, rounded to protect the linen, was turned by a handle to wash the clothes and then the dirty water drained away through a tap into the sink.

In 1870, in Pittsburgh, USA, Anna Smith's 'Lady Washing Machine'[25] was basically two washboards placed horizontally in a wooden tub and activated against each other. By 1907, a Berliner, Elisabeth Beckmann, made her improvements to 'Cylinder Washing Machines'.[26] No matter in which direction the rotating drum turned, water would be scooped or sucked out accordingly: 'one chamber

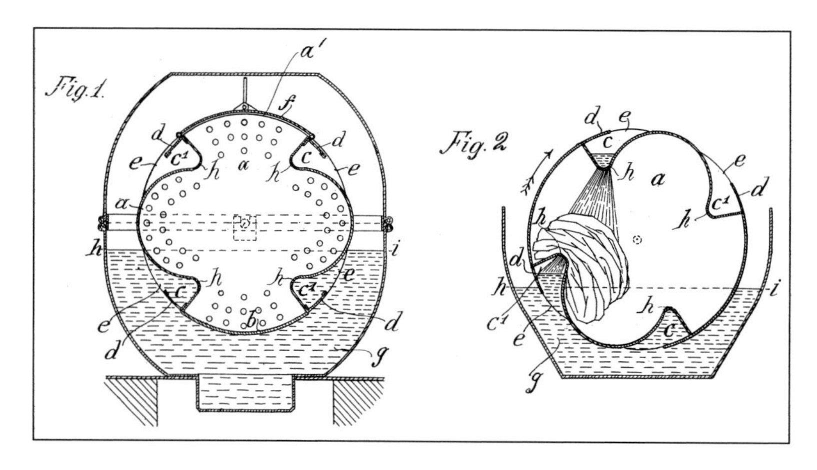

Dirty clothes were washed in the rotating drum of Elisabeth Beckmann's machine. Patent.

scoops up water and allows it to trickle down through its perforated wall upon the clothes or other linen in the interior of the drum . . . the following chamber, owing to the vacuum arising in it, sucks the water out of the clothes lying in the drum'. This was an early automatic washing machine. There were even attempts to combine a machine to wash clothes and dishes. Clothes drying was considered by Frances Florinda Crawford from Ontario with her 'Apparatus for Preventing the Shrinking of Garments During Drying after Washing',[27] especially for cotton and woollen clothing. She made a flat frame in roughly the dimensions of the torso and two hinged arm shapes, which could be easily folded for transportation; wet garments were pulled over the frame to dry.

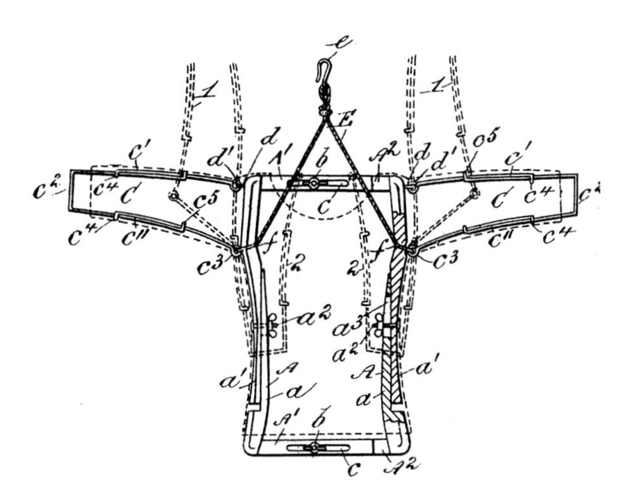

Shrinkage of woollen garments could be prevented by using Frances Crawford's frame. Patent.

Washing Up

Progress on mechanising the washing-up of dishes follows a similar path to that of the laundry. The goal to automate it was finally achieved by Josephine Cochran from Chicago in 1886.[28] Cochran, née Garis, was the daughter of a civil engineer and had married a man who hit hard times and then died. She in the meantime was determined to perfect her machine to wash dishes and, when she eventually did so, established the Garis-Cochran Dish Washing Machine Co. Her machine was exhibited at the Chicago Exhibition in 1893 and was sold to the restaurants opening in the new, confident city.[29] The Garis-Cochran dishwashing machine consisted of a chamber in the bottom of which cylinders of water and soap were placed and above them a handle which turned ratchets above it, that in turn activated a pump in the water or soap cylinder. A crate containing the crockery rotated to catch the soap and

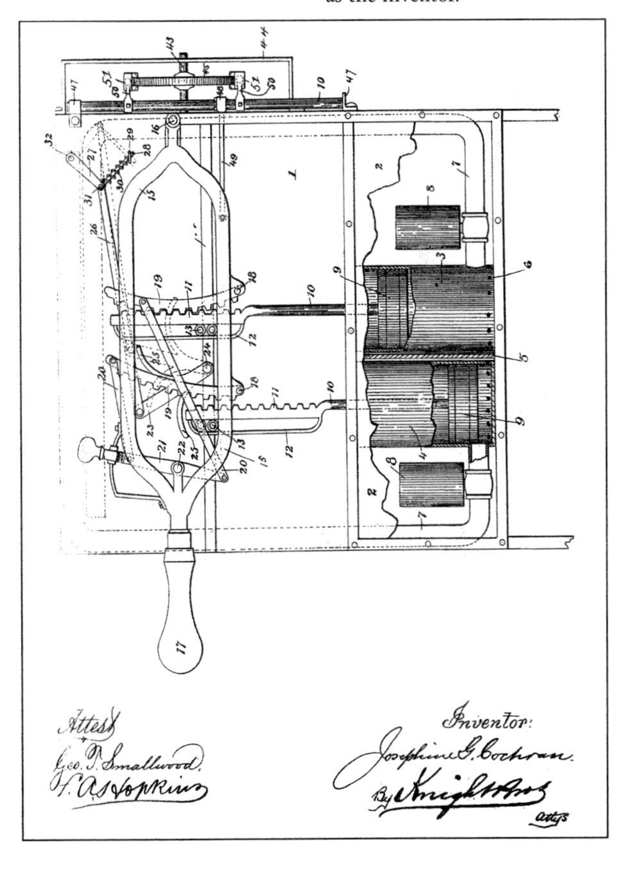

Josephine Cochran's patent for her dishwashing machine in which she clearly signs herself as the inventor.

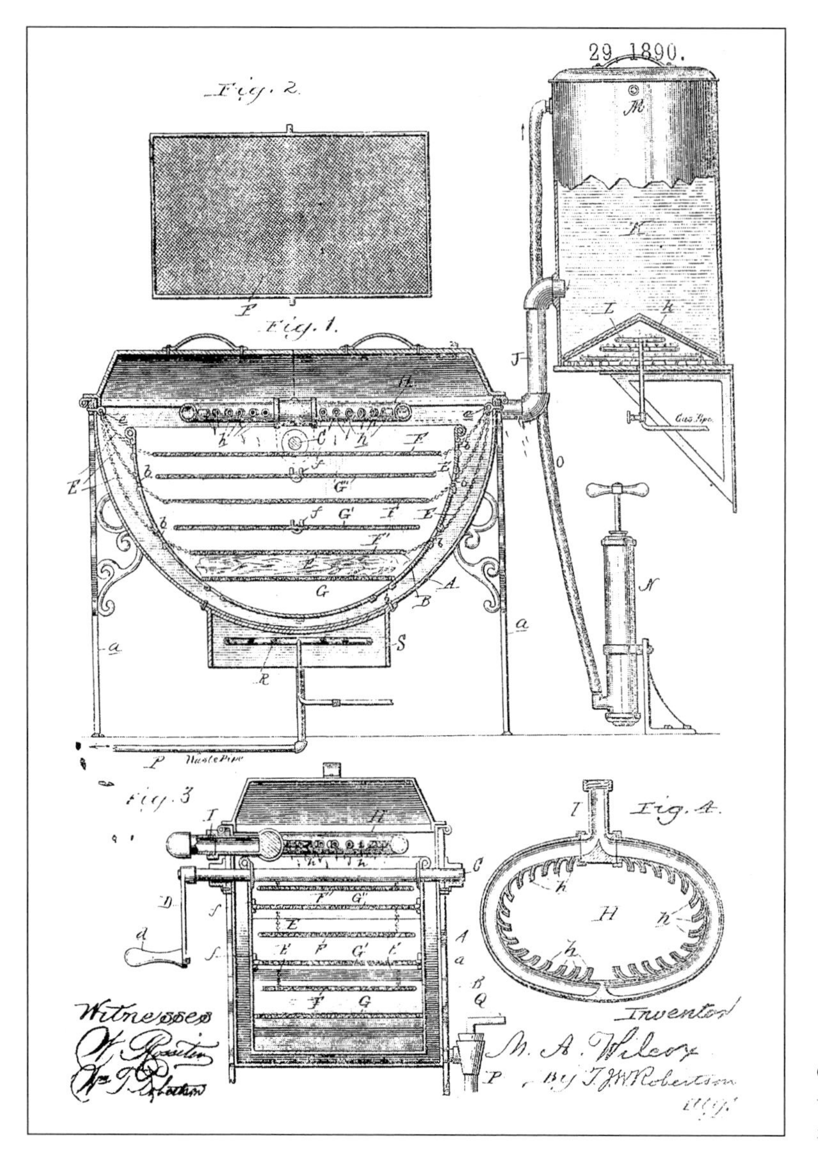

Clothes and dishes could be washed in Margaret Wilcox's machine. Patent.

water as well as drain and dry the pots. Josephine Cochran claimed that 'a continuous stream of either soap-suds or clear hot water is supplied to a crate holding the racks or cages containing the dishes while the crate is rotated so as to bring the greater portion thereof under the action of the water'.

Around the same time Margaret Wilcox, also from Chicago, designed a machine to combine the washing of dishes and clothes.[30] This semicircular tank could be supported on ornamental legs or sit on the kitchen draining board. Inside it a crank rocked a cradle. Rubber mats acted as a washboard for the clothes and were removed when the dishes were to be placed in it. A gas boiler, attached to the mains water supply, filled the tank with hot water. In 1908 Fanny Flower,[31] a married woman from Pendleton, was concerned at the detrimental effect to the hands from washing up experienced by 'persons who in consequence of a limited income are unable to retain a general servant'. Instead of wearing a pair of India-rubber gloves she fixed a series of grips and frames to the end of long handles, which were attached to dishes, cups and plates. The crockery was dipped into the water and a washing up mop, on a short stick, in the other hand did the washing. In this way neither hand got wet. Isabella Peckover's triangular 'Sanitary Sink Basket',[32] was placed in the corner of the basin to collect all the detritus of washing up and to stop pipes becoming blocked.

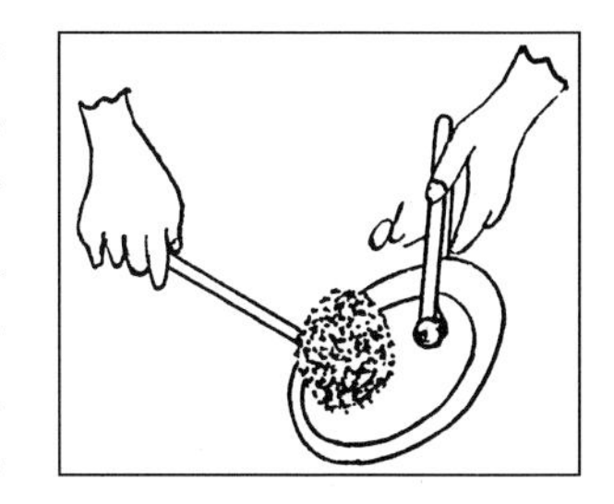

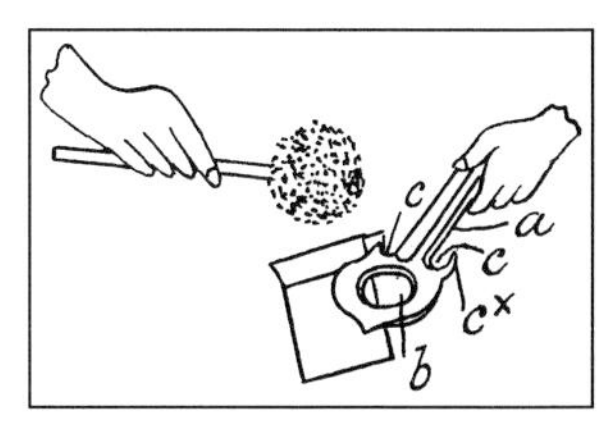

Hands were kept dry with Fanny Flower's long-handled dishwashing devices. Patent.

Food and Cooking

Although gas was increasingly used in the home for lighting, its use to heat cooking stoves was slow to spread. The chef Alexis Soyer did install gas stoves at the Reform Club and they were exhibited at the International Exhibition in London in 1862. But there was public prejudice against them partly because the emission of gas, through the jets, was impossible to regulate and thus likely to ruin the cooking. For much of the nineteenth century stoves were heated by solid fuel and it was not until the First World War that gas-powered cooking appliances became widely available.[33]

Some women, like Amelia Lewis, devised small portable cooking stoves for domestic use and larger ones for the army. The philanthropic and entrepreneurial Mrs Lewis, anxious that the poor should eat nourishing food and keep warm, designed her small domestic stove to run on peat. Her accompanying booklets advocated the beneficial effects of steaming food and alcoholic abstinence. In 1869 Elizabeth Woolcott Slade and Maximelia Slade from Wilton near Salisbury designed an 'Improved portable Cooking Apparatus'.[34] Standing on legs, this semicircular, portable oven was placed in front of an open fire. It consisted of a series of ovens, each lined with metal to reflect the heat, which could be built up in layers, one upon another. Each section had a movable screen to regulate the amount of heat, shelves within each oven and a dripping pan. The top oven had a lid and could be used for cooking smaller items like chops.

Marie Agnes Auzeric from Paris patented her cooking stove in Britain in 1903.[35] This was a gas stove of enamelled iron with burners at the base of the oven and beneath the hob. To alleviate the perennial problem of the conventional gas stove, where the gas burners were too close to the cooking food and their fumes affected

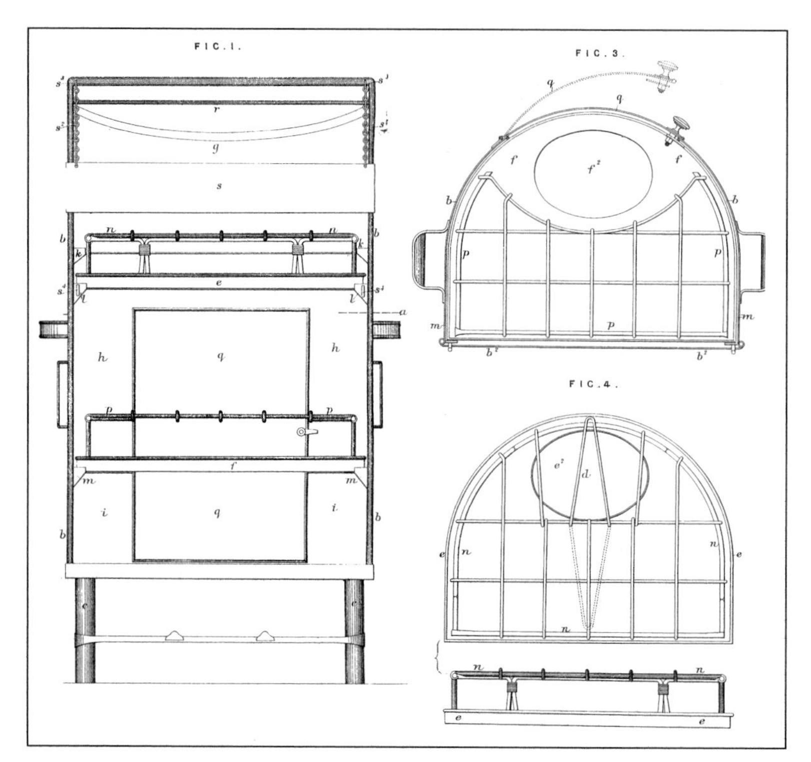

The semi-circular, portable stove designed by Elizabeth and Maximelia Slade. Patent.

the taste, Mme Auzeric proposed several ovens. The food sat on grills with a basin beneath to catch the juices and each oven was insulated from the burners which could be fixed at the bottom or moved around depending upon the position of the oven.

The removal of hot dishes and plates from ovens required attention and Frances Tenison, a 'lady' and her husband Adolf, a medical student of Shepherd's' Bush, London[36] devised a frame with clips around its edges and wooden handles at one end. When placed beneath the oven shelf and the handles depressed, the clips were activated and attached to the shelf which could then be pulled out without the user getting burned or the contents falling on her.

The large household inevitably meant a lot of cooking and, although Katy Liddle's patent application gives an insight into the domestic chores of an Edinburgh cook and housekeeper in 1863.[37] Each morning her master requested a cooked egg for himself and guests. In order to speed up the process she had a double-bottomed saucepan made, the upper level having holes in it, large enough to take a 'small tea cup'. These cups were then held in place and immersed in the boiling water beneath; the eggs were cracked into them and left to cook. Proudly, she described her master's reaction to her method of coddling eggs:

> He says that it very much improves the flavour as compared with the ordinary method of boiling an egg in the shell; it is also so nice and clean looking, and you can also stand by it always and at once detect an unsound or imperfect egg. . . . (He) is at pains to shew it off to any friend . . . who may be visiting him, whether it be at breakfast, or at dinner, or at supper, by having an egg cooked . . . He likewise says that it makes the egg so light and easy of digestion, that he thinks a man might with ease eat half a dozen at a meal without any injurious consequences.

Two women coffee makers, Sarah Guppy and Melitta Bentz, devised ingenious methods for its preparation. Sarah Guppy was married to Samuel, a Bristol merchant. She was an intriguing woman, certainly undaunted by the differences in scale of her inventions for tea and coffee urns, beds and bridges and railways. In 1811 she patented her method to make the piling safe for a suspension bridge, some years before Thomas Telford came up with his design for a bridge at the Menai Strait, and the following year she patented her 'Improvements in Tea and Coffee Urns'.[38] This was a multifunctional and highly practical urn; not only did it make tea and coffee, it cooked eggs and kept the toast warm. An oil lamp or candle heated the urn full of water, then a 'small vessel' was placed or suspended in the top of it so that 'it . . . shall receive heat from contact or immersion in the heated fluid . . . and be capable of receiving one or more eggs and of boiling or cooking the same . . .'. On the top of the urn she placed 'an elegant and convenient support for a plate or small dish or other vessels to contain toast or other article of food . . . and by contact with the heated fluid or the steam or vapour proceeding there from to keep the same well and sufficiently heated'.

The innovation of a German housewife, Melitta Bentz from Leipzig, formed the foundation of a huge international company, which still produces a development of her original idea. Like many of the best ideas, this was very simple; her solution to the problem of the nasty taste produced by the dregs in the bottom of the cup was to find a way to filter the coffee. She rolled a piece of blotting paper into a cone shape, and put it into a brass pot, into the base of which she had drilled holes; this was then placed on a jug or coffee pot. The ground coffee was placed in the paper cone and hot water poured through the first filter paper and the Melitta brand was born. Her ownership of the idea was registered in Berlin in 1908 as a 'Gebranchsmuster' (Utility Model).[39] Unfortunately, it no longer survives, having been destroyed in the Allied bombing of Berlin in the Second World War. But the Melitta brand name, synonymous with strong, fresh, hot coffee, is known throughout the world.[40]

Melitta® trademark.

Melitta Bentz, inventor of the Melitta® coffee filter paper. (*Melitta Haushaltsprodukte GmbH & Co. KG*)

The need to extend and preserve the life of fresh food and spices is very old; with this aim, meats, vegetables and spices had been dried, smoked, marinated and salted to provide nourishing and plentiful food through the long winter months. The search for new and more successful methods of preservation has also been long and experiments not always easy, straightforward or successful. Some have led to food poisoning and contamination, especially when hygiene practices and the toxicity of materials were poorly understood. Illness and death resulting from a new method of preservation were not uncommon. When Amye Everard als Ball became the first woman to take out a patent in 1637,[41] she was part of this tradition. Her patent was for the preparation of a tincture of saffron, roses, etc., which she had developed by 'her own invention and industry, and by infallible experience, at her own charge'. Saffron, the precious stamens from the crocus flower, with its rich golden hue and strong aroma, was used to flavour the dishes of the wealthy and still retains an exotic status among other herbs and spices. Mrs Ball's method was to make it easier to use than previously and preparing it in such a way that when added to a tincture it would retain the full strength of the saffron for 'many years'. Her method could also be applied to 'roses and gilly flowers . . . rosemarie flowers, coweslipps, and lavendar flowers'.

Elizabeth Burgess inherited, with stepsons, her late husband's business making Burgess's Anchovy Essence and in 1859 took out a patent, in her name, to retain ownership of the method of preparation in the light of competition from rivals.[42] Anchovy essence was more of a paste than an essence and much enjoyed by officers in the navy who took consignments of it in porcelain, lidded pots on their long voyages. The nineteenth century saw an increase in sea travel to and from the new settlements in the colonies. This fuelled the desire to receive perishable food from faraway lands, especially meat and fruit, so new techniques for food preservation, other than the traditional salting, pickling or drying were required. There were many attempts at tinned food with ghastly tales of botulism and similar deadly diseases resulting from its use. Refrigerated ships were introduced in the 1880s to bring huge consignments of beef and lamb from South America, New Zealand and Australia. On a smaller scale, Amanda Jones, an inventor from Wisconsin in the United States, found a method to vacuum pack fruit in jars.[43] By clamping the convex lid, lined with cord, on top of flanges on the jar, then placing a large packing ring on top, air was extracted through a valve by turning it though 90 degrees. Amanda Jones and her cousin were, individually, awarded numerous patents for their preservation methods and she founded their Women's Canning and Preserving Co. in Chicago which advertised its puddings, lunch tongues and fruit preserves. Later she ventured into the Pennsylvanian oilfields and showed the oil workers how to avoid getting burnt using a safety valve to control the amount of crude oil flow to be burned.[44]

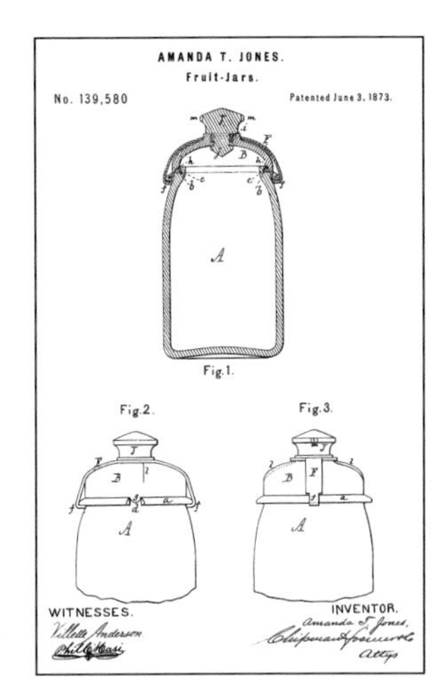

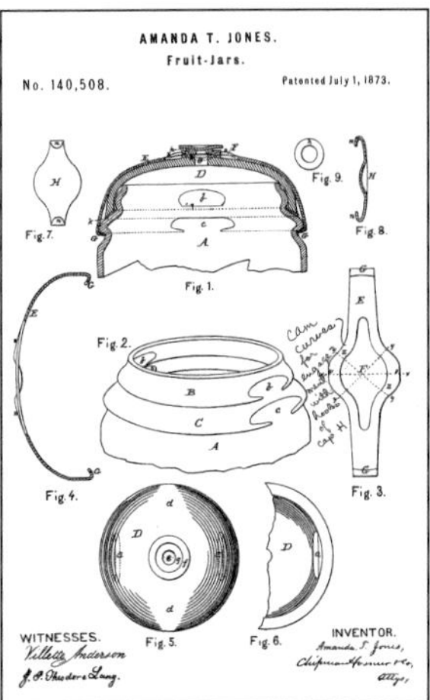

Amanda Jones's vacuum-sealed food preservation jars. Patent.

Maud Walkey[45] was concerned at unhygienic practices in the scalding of milk in order to preserve it. To a tank of water she added chlorine or calcium, chemicals with a higher boiling point than water. The milk passed through a metal coil which zigzagged through the boiling water and chemical and could be removed easily for cleaning. Farms provided a fertile ground for invention as demand for produce in more distant markets grew. In 1904, Theodora Wilson-Wilson from Westmoreland made oval, artificial 'brooders' for chickens[46] specifying that each was used only for small numbers of chickens to prevent overcrowding.

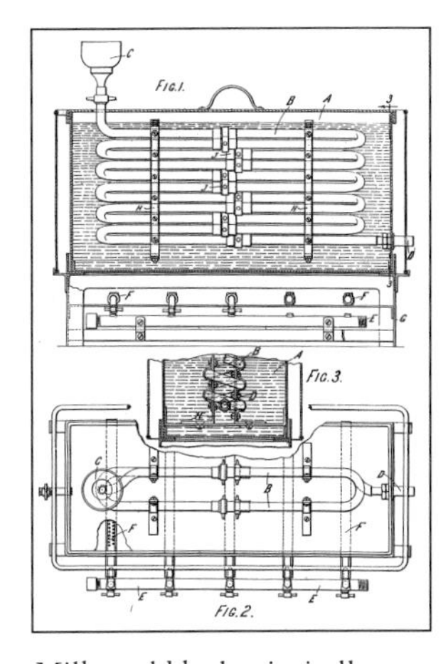

Milk could be hygienically scalded and preserved in Maud Walkey's machine. Patent.

Even bottles and stoppers excited attention. In the era before the plastic bottle the problems encountered with broken glass were enormous. In 1864, Elizabeth Brimson from Frome in Somerset[47] designed some very elegant bottle covers made from paper, cotton, straw reed or rush glued together. Another example, by Elsie Josephson from East Molesey[48], 'prevented fraud' whereby a bottle could be tampered with and an unfavourable liquid poured in. She proposed an indentation on the inside and outside of the rim of the bottle and placing a wire within it. When filled and corked, if then opened, the bottle would break on the weakened line of indentation making any tampering obvious. Mrs Josephson was heedless of any danger from consuming shards of glass, as she recommended corking the broken bottle and using the contents at a later stage. In the same year a mineral water manufacturer from the Isle of Wight, Julia Richards, suggested engraving or embossing advertisements on the inside tube of soda water siphons.[49]

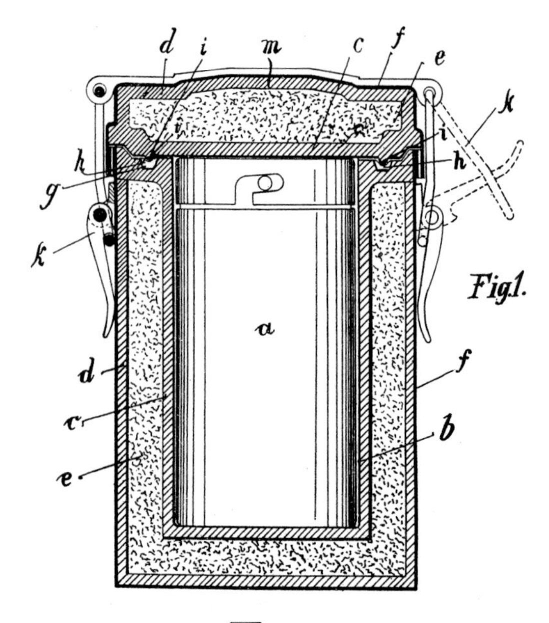

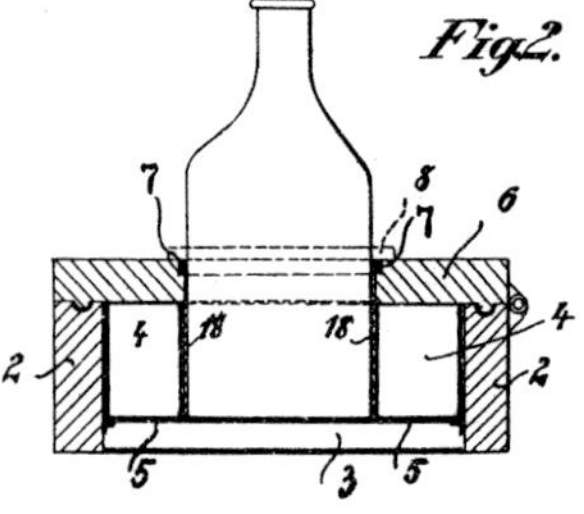

Dora Buhlman's heat insulating device for both hot and cold foods. Patent.

Ice had been imported from the Arctic for use in the icehouses of the very wealthy, but the new technologies brought a plethora of ideas to make refrigerators for domestic use. Agnes Marshall claimed to have invented the first ice-cream maker, although the patents were probably taken out in her husband's name, but she herself wrote two books *The Book of Ices* and *Fancy Ices*.[50] Dora Buhlman, of Kaiserslautern in Germany, devised a dual-purpose 'Heat Insulating Device'.[51] She cites the failure, due to faulty closing devices, to keep the contents at the correct temperature as need for improvement. In this machine the vessel is hermetically sealed allowing no steam to escape from the hot contents and no warm air to enter when used as a refrigerator. Again, any understanding of food poisoning was minimal, when, in its use as a digester, hot, partially cooked food was placed in the vessel and sealed so that it would continue cooking and remain hot until heated. The same principle with a reverse effect took place when ice was placed in it. The digester was made of two compressed cellulose or wooden walls, forming a hollow cylinder, which contained a non-conductive material, like waste silk. The lid was held securely in place with a tight metal fastener.

Kitchen Equipment

A precursor of the modern fitted kitchen was the kitchen cabinet from the United States. There women had designed a piece of furniture to contain all the kitchen equipment they needed and to be stored in one place. Homemade at first, these cabinets eventually became essential pieces of manufactured furniture by 1899.[52] In 1908, Ella McAllister Bennett from Oklahoma introduced her kitchen cabinet to Britain.[53] Part dresser, part cupboard, it had

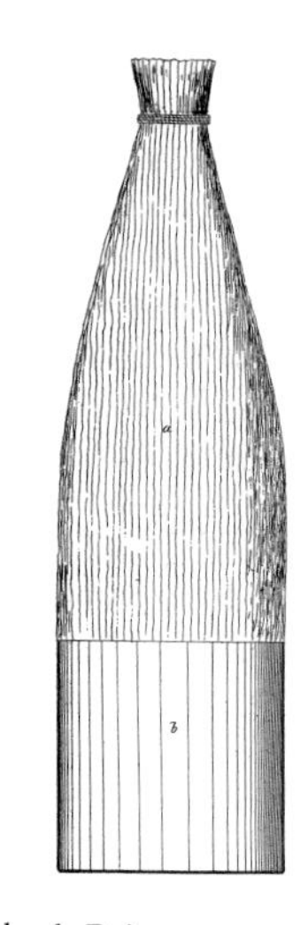

Elizabeth Brimson covered glass bottles to protect them from breakage. Patent.

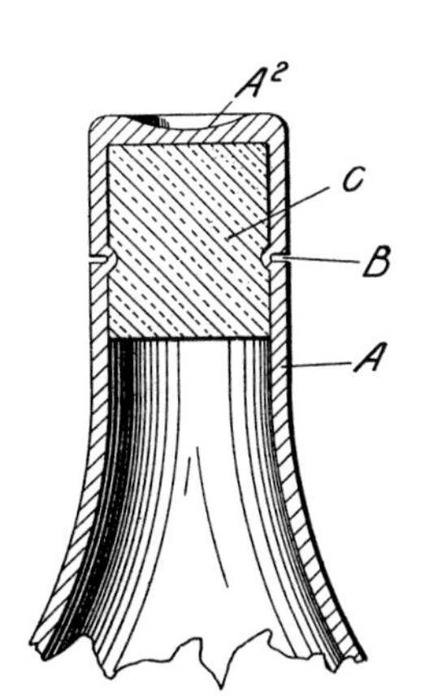

Tamper-proof bottle stoppers by Elsie Josephson. Patent.

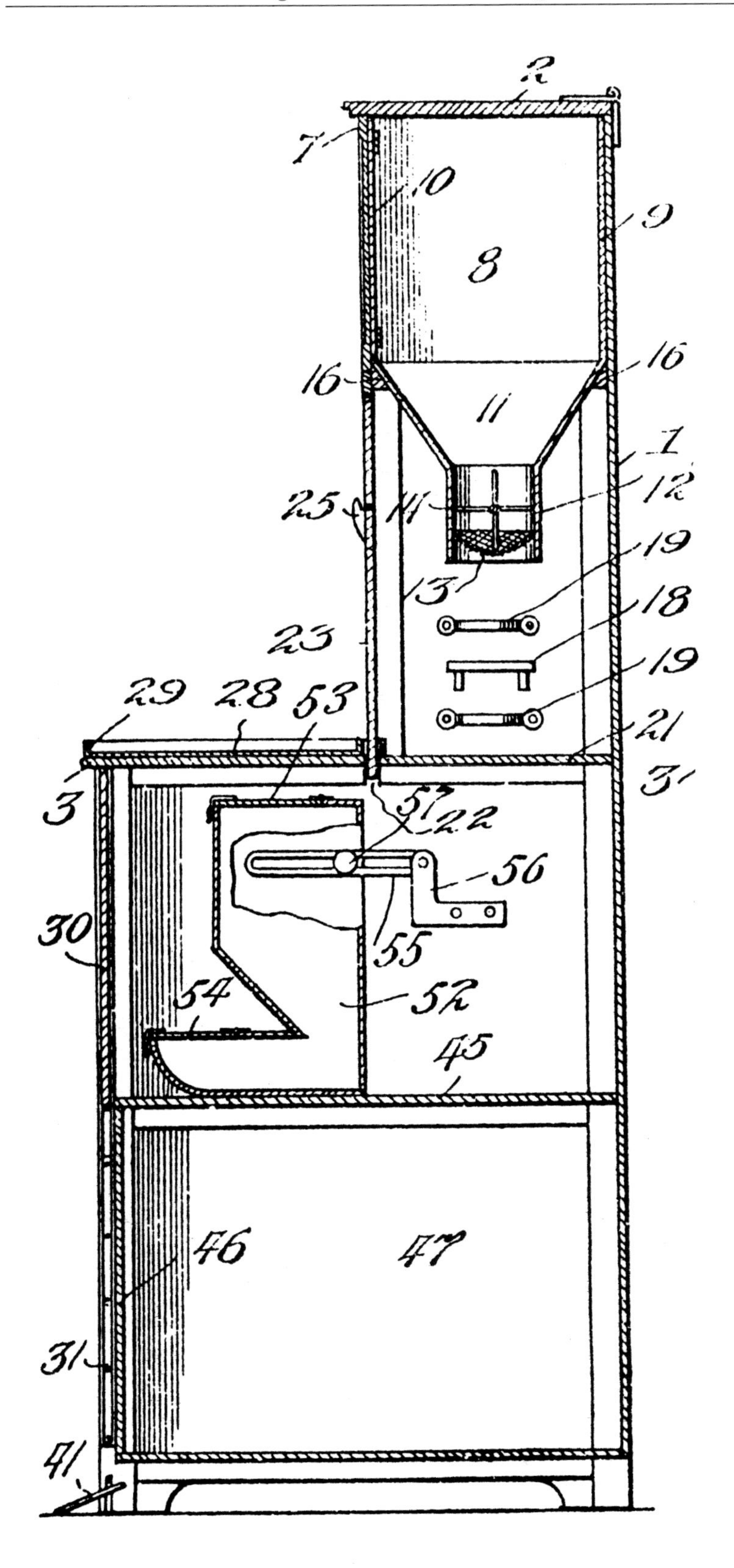

The integral kitchen cabinet, like this one patented by Ella McAllister Bennett, was popular in the United States. Patent.

sliding doors operated by a foot pedal; a sugar bin mounted on a sliding, pivoted arm; a metal flour bin with a hinged door and discharge nozzle; a table which could be pulled out from the base of the upper part of the unit; and hooks and shelves for spice bottles and utensils as well as storage cupboards.

In Britain, there was still concern about protecting the hands, this time from being burnt by hot dishes. A solution was proposed by Clara Ann Tyler from Birmingham,[54] a city renowned for its small metal goods industries. She made a series of dish and plate holders from two metal clips attached to a spring bow similar to a pair of scissors. Upon pressing the spring bow the clips would grasp the dish thus allowing it to be picked up. Mrs Tyler designed a series of different shaped clips to make decorative additions to the gadget. Clearly for a middle-class household, eager to impress with its dinner parties, she states that by using the holders 'it [is] unnecessary for servants or others to touch the plates, dishes, and like articles with the hand at table, and also preventing the burning of the fingers by heated plates and dishes'. The minor but real problem of grazing the fingers when grating a nutmeg was attacked by Emma Thurgood from Bayswater and Leonard Haselwood, a P&O clerk, with their nutmeg grater in 1905.[55] A small, movable box was attached to the grater and the nutmeg placed in it. By holding its sides the box could be pushed up and down the grater and the 'troughs lying between the rasping surface and the thumb and fingers, prevents the same from being sacrificed by the said rasping surface'.

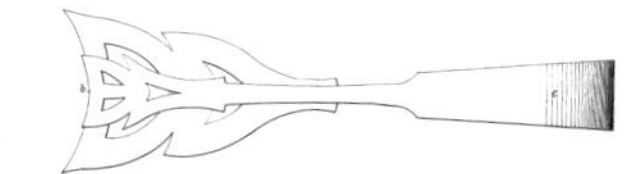

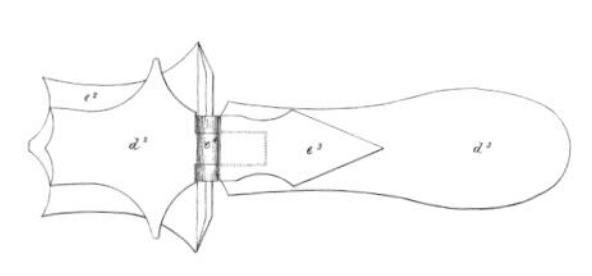

Some of Clara Tyler's metal dish and plate holders. Patent.

Even the standard pudding bowl did not escape modification by Mrs Anna Lawson from Dublin[56] who added a projecting band ¼ inch beneath the rim of the bowl on the outside. This beautifully simple solution enabled the string holding a cloth cover over the bowl to stay in place during steaming. But three years later Ada Anna Ashmall-Salt from Buxton,[57] eradicated the bowl for steaming and the problem of the pudding sticking to the cloth cover, with her 'pudding mould primarily to steam or boil roll jam puddings.' Hers was a metallic or earthenware mould, perforated and divided in two parts with a hinge joint running down one side and a clasp on the other. There were feet on the bottom to stand in the water of the steamer. When cooked, the pudding would roll out in one, perfect piece. By 1901, with yet more technological advances making their way into the home, Bertha Wells from Brighton designed a piston-operated mustard pot[58] to 'deliver the condiment directly upon the plate without the use of a spoon'. A plunger forced the mustard out through a hole in the lid.

Luise Kopmann from Cassel in Germany[59] advocated making a hearty soup in her enamel iron or aluminium saucepan and 'sieve'. By placing meat, bones and seasoning in the sieve and then into the pan

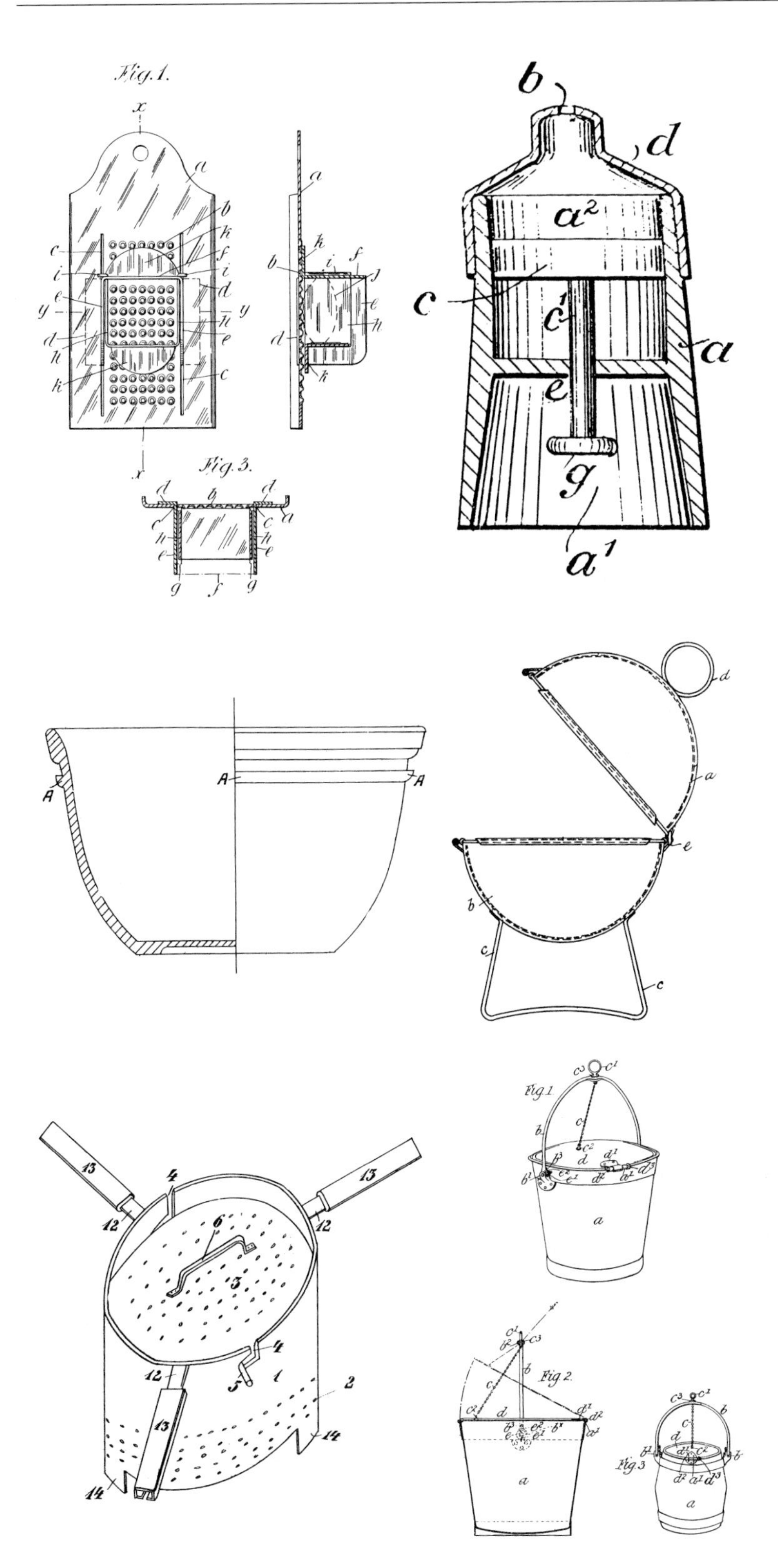

Left to right from top left: Nutmeg grater, Emma Thurgood. Patent, Piston-operated mustard pot, Bertha Wells. Patent. Improved pudding bowl, Anna Lawson. Patent, Pudding steamer, Ada Ashmall-Salt. Patent. Saucepan and sieve for soup making, Luise Kopmann. Patent. Chain-activated pan or bucket lid, Edith Marshall. Patent.

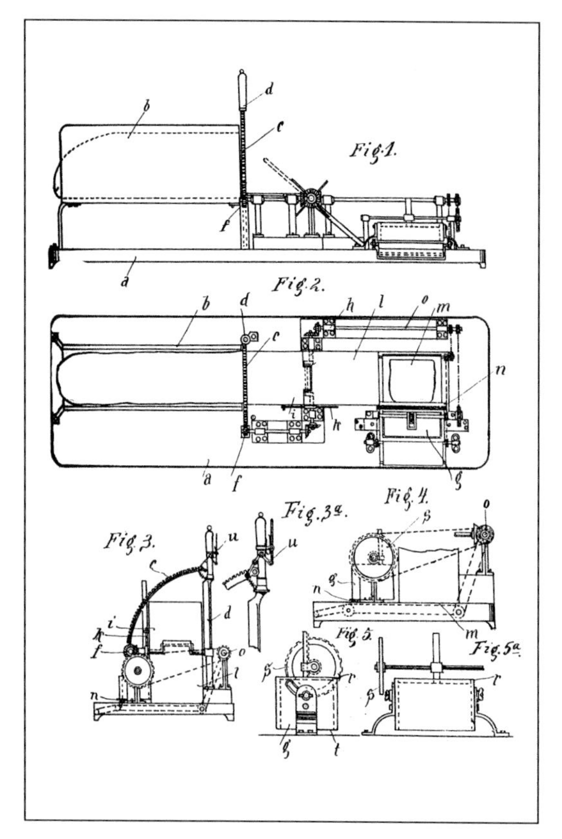

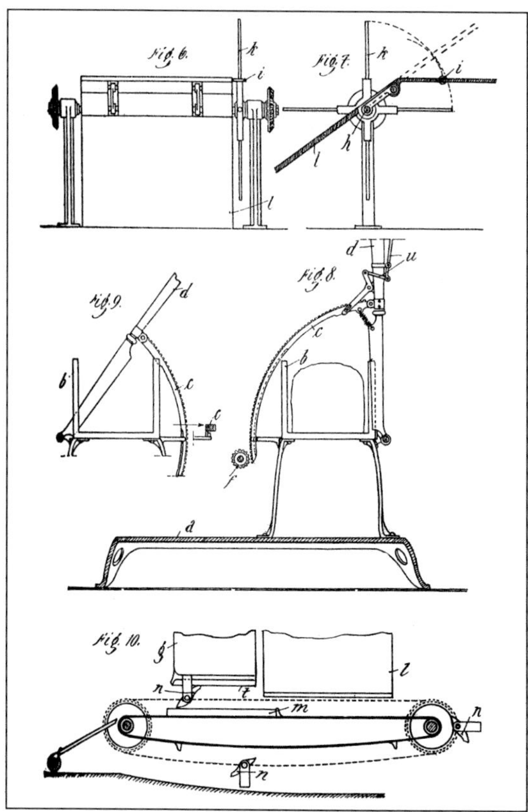

Bertha Paatsch's bread slicer and butterer. Patent.

of water, it could be taken out without having to drain it further, leaving a strong, clear stock behind. Edith Marshall[60] from Chelsea, London devised a saucepan with a lid attached by a chain to the handle. By moving the handle the lid was activated and opened. She thought it would be useful to a host of people for a variety of purposes including saucepans, boilers, steamers, portable or travelling baths, sanitary and hospital utensils, and vessels used in photography. One woman, and undoubtedly there were many more, became overexcited at the prospect of registering her inventions. Augusta Barnett from Highbury in North London, and later Essex, applied for seventeen patents between 1903 and 1907. All were for small, almost incidental items, and only one, an improved tea strainer, received a full patent.[61] This consisted of a wire coil with a conical lower end, which was placed in the spout and hooked on to the rim of the spout.

By 1910 the number of applications for British patents made by German women was increasing and the majority of them, even for household gadgets, were based on a technical and engineering approach. Two, by housewives, were for bread slicers: one by Elsie Schmidt[62] and the other from Bertha Paatsch.[63] The latter's machine also buttered the bread. Through a series of gears, axles, toothed racks, flaps, cranks and chutes the bread is sliced. Redolent of a factory assembly line transplanted to the domestic kitchen, the machine cut the bread and passed the slice through a flap on to a chute; the flap closed to catch the next cut slice, while the first continued its journey on to the buttering device:

> Here the knives *n*, which are fixed on endless chains and driven from the toothed *c*, *f* through an axle *o*, pass directly above the slice *m*, after they are previously led along the butter receptacle *g* and provided with a sufficient quantity of butter, whereby the slice *m* is buttered. The chains for the knives are further fitted with abutments which move the buttered slice from the buttering device before the next slice arrives thereon.

This 'Improved Machine for Cutting Bread in Slices and Buttering the Same' is a fine example of trying to better a perfect and simple tool, the knife.

Designing the Home

With the expansion of town and city boundaries by the mid-nineteenth century came a proliferation of housing: endless two up, two down terraces for the working classes, enormous mansions and substantial villas for the middle classes and the country retreats of the mill owners and merchants. A distinctive Victorian style developed, much of it influenced by the preference of the designs of the houses of the Queen and her husband Prince Albert. Despite her forty years of reclusive widowhood that began in 1862, Queen Victoria had, in earlier years, presented her people with a visual image of the 'ideal family'. Around the world lakes, rivers, towns, roads, streets, houses, pubs and institutions were named after Victoria, Albert and their children. Two distinct and diverse historical styles influenced the architect, reflections of both of which can be seen in domestic design. One was the desire to recreate the grand Gothic tradition with its pointed arched windows, redolent of the great cathedrals of England. This was used by both George Gilbert Scott in his great cathedral to the steam engine, St Pancras station with the Midland Grand Hotel in London, and by Alfred Waterhouse in his cathedral-like Manchester Town Hall. The other was the classical, based on the Greek traditions of architecture, with its system of columns and pediments, which was used in the design of countless institutions of learning and greatly influenced the first combined architect/builder Thomas Cubitt. He developed areas of London

The Midland Grand Hotel at St Pancras designed by George Gilbert Scott. (*Jaqueline Mitchell*)

such as Belgravia, with its large, distinctive white terraces and squares with columned porticoes and rectangular windows, and his influence spread out to the numerous white stucco villas in the cities and towns throughout Britain.

Indoors, the Victorian approach tended to be a hotch-potch of style, among their furniture, gadgetry and machines. All of these might be highly decorated, revealing a love of curved lines and, at times, exotic images which disguised the purity and simplicity of function. In addition to the gothic and classical influences there were colonial imports from the cultures of India and China. The Victorian attitude to design and style is well described by Nikolaus Pevsner (1902–83) in his critique of the exhibits shown in 1851.

> A universal replacement of the straight line by the curve is one of the chief characteristics of mid-Victorian design. As against other styles favouring curves, the Victorian curve is generous, full or . . . bulgy. It represents, and appealed to, a prospering, well-fed, self-confident class. A peculiar top-heaviness often to be met is only a special case of this delight in abundant protuberances . . . There must be decoration in the flat or in thick relief all over all available surfaces. This obviously enhances the effect of wealth.[64]

Among her other skills, Mrs Beeton advised on planning and arranging rooms and displaying artefacts and souvenirs from travels abroad. Gradually, the Victorian home became ever more flamboyant and by the 1840s the love of colour and embellishments was already evident. In 1839, an artist, Marie Françoise Catherine Doetzer Corbaux, living in London, developed paints, with her colleagues Francis Spilsbury, a chemist and Alexander Byrne,[65] which could be easily washed when applied to walls, or even used by the artist. Previously pigments were mixed with oils, gelatine or size and then applied to walls or papers but were not colour-fast when washed with soap and water. By mixing their pigments with specific chemicals like borax and shellac they were rendered colour-fast when washed and even harder wearing. Mary and Richard Cole of Bayswater lavishly decorated their windows and furniture and appreciated a love of colour and decoration.[66] By intermixing colours they stencilled, printed and painted or used transfers to discover that 'in some instances it may be found desirable to emblazon such coloured devices with gold, silver, bronze, or other suitable metals, or a device may be placed upon the glass and a plain colour either filled into the interstices or laid over the entire back, all of which devices showing through the glass in front thereof'.

Reminiscent of Mrs Coade and her stone was Elizabeth Wallace of Cheltenham[67] who in 1848 patented a new, artificial, yet decorative, stone and a complete scheme to decorate the exterior and interior of a house. Specifically, she claimed, it could be used for

> the walls, porches, pillars, pilasters, and other external parts of houses and other buildings with a combination of materials which is less liable to be affected by wet, damp, and atmospheric impurities than any material . . . hitherto in use for this purpose, is of a beautiful lustre and great durability, and may be able to exhibit externally any color or mixture of colors or disposition of colors, arbitrary or natural, and also to resemble closely any natural building material . . . for example, pure white marble, veined marble, shell marble, porphyry, malachite, granite &c.

The stone was made in any shape or size and composed of an outer sheet of glass beneath which was a layer of wet plaster of Paris, which might be coloured with pigment to resemble one of the stones to be imitated. When dry, the glass was removed and the rough edges of the plaster coated with white lead paint and screwed into place on a wall of wet cement. For interiors she added decoration: landscapes, figures, fossils and plants engraved into the glass, or another surface, to make a relief in the plaster. She detailed how to make letters and incorporate them into the scheme and all could be painted and embellished with gold leaf or paint. A large stencil diagram of a

One of Elizabeth Wallace's decorative designs for her stone-like substance. Patent.

grieving maiden resting on an urn with the words 'In sacred memory' beneath was included. In 1851, calling herself an inventor and now based in Fitzroy Square in London, Elizabeth Wallace exhibited her slabs of imitation marble at the Great Exhibition. She claimed that by following the patented process the glass could be as solid as stone and easier to repair in case of accident.

The indomitable and ingenious Sarah Guppy, sadly widowed by 1831 and living in Clifton, Bristol, came up with her third invention for a splendid metal-framed bed, with curtains around it and an

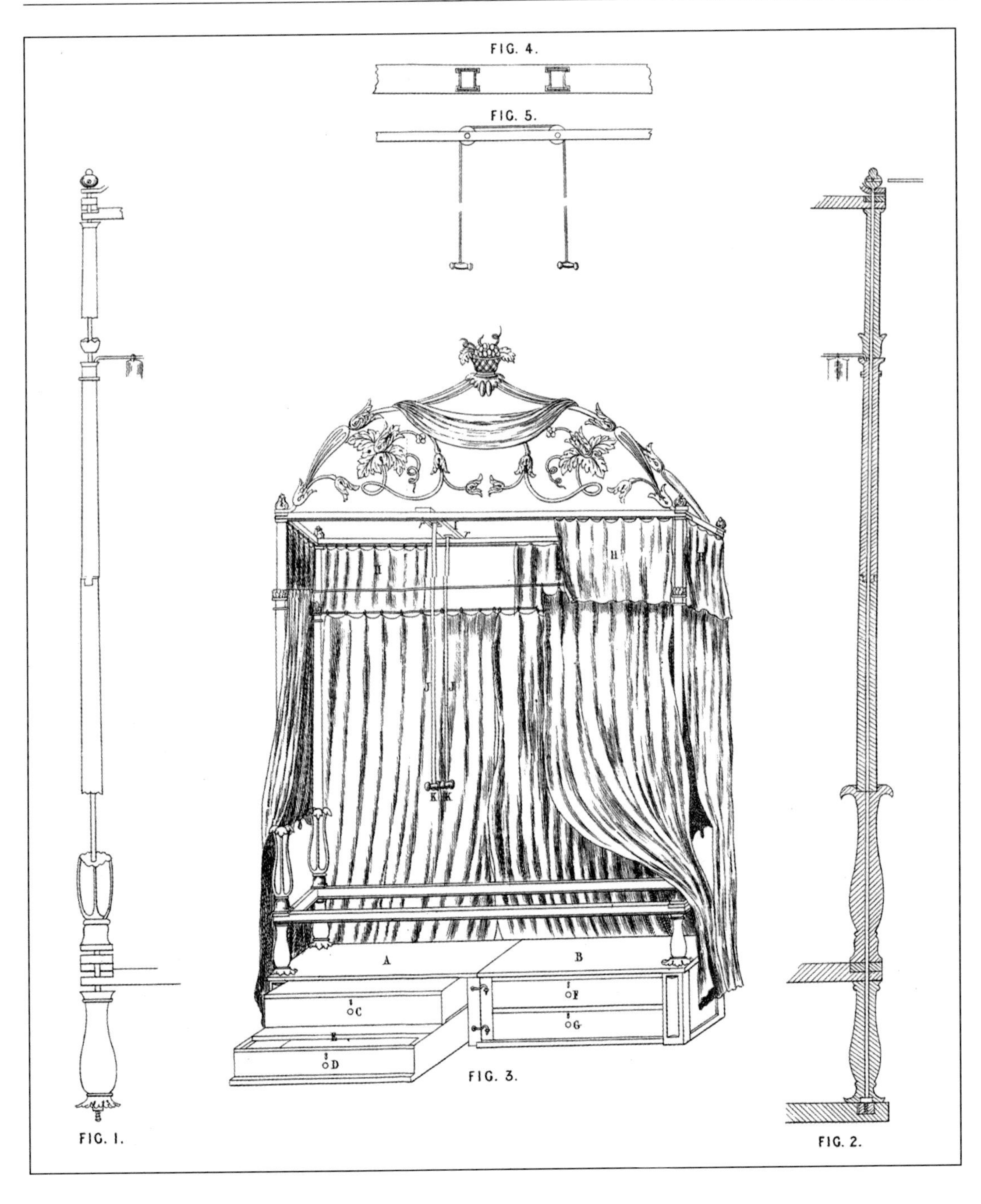

ornate top.[68] What made this bed different, apart from its metal frame, were the drawers beneath it and the exercise bar suspended from a beam across the top of the frame. When opened, each drawer had a sliding lid which formed a step when closed; thus the user would climb the steps into bed, draw the curtains to keep warm and exercise on the bar operated on a pulley system.

Sarah Guppy climbed into her bed and exercised from the bar suspended above her. Patent.

By contrast, some of the inventions by women in the nineteenth century were for simple, utilitarian items of furniture, possibly designed to alleviate crowded conditions or for travelling. In the mid-century, at a time when women were encumbered by their crinolines, a widow, Louisa Monzani, won the patent rights of her deceased husband's estate[69] for beautiful and refreshingly simple furniture designs. Mrs Monzani lived in Old Kent Road in Surrey (now south London), her husband having run his business in nearby Bermondsey. The patents are for a folding bedstead and a chair. The first, for 'Bedsteads and Packing Cases or Boxes to Contain the Same and Other Articles', in which 'this invention is so to arrange parts that a bedstead shall be partly constructed of the box or case which, when the bedstead is packed and out of use, contains it'. Constructed in two parts, it consisted of one part fitted out like a portmanteau to carry clothes (a wardrobe) and the other to hold the bed and bedding. When pulled apart, strong sacking and straps were stretched over them to form a bed. Stretcher bars were applied down either side to give extra stability. One end of the box could be adapted to form a bedhead to stop the pillow falling off and the whole thing was packed away into a box ready for transport. Her second patent was for a folding, wooden-framed chair, with a canvas seat, the simplicity of which would not be out of place in any modern furniture store. Later, a simple, collapsible table would be used as a games table by Catherine Dowie and Charlotte Easson from Glasgow.[70] The hinged legs could be folded up, and the separate tabletop made of laths or slats of wood could be rolled up. The top was covered in green baize for card games on one side and the slats on the other were ruled into squares as a chess or draughts board.

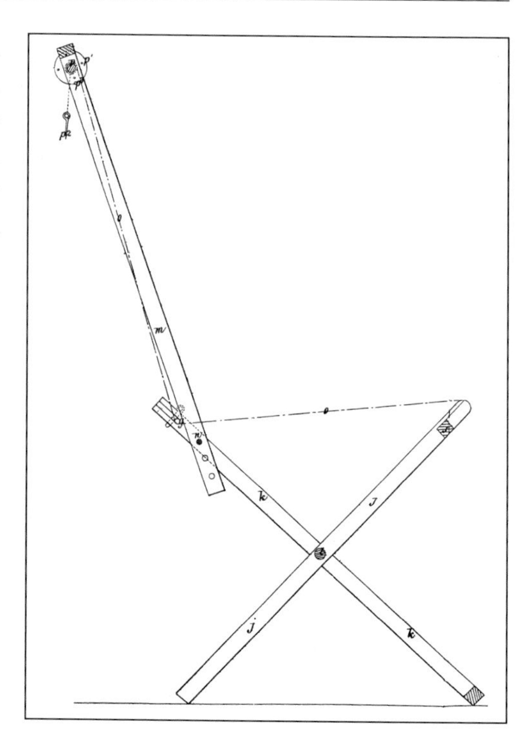

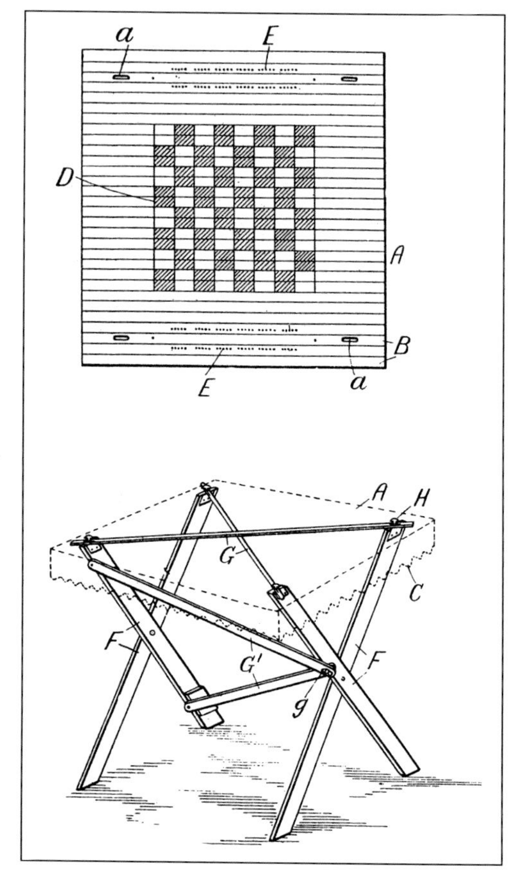

Above, right: The simplicity and functionality of Louisa Monzani's chair were ahead of the times. Patent.

Right: The collapsible games table designed by Catherine Dowie and Charlotte Easson. Patent.

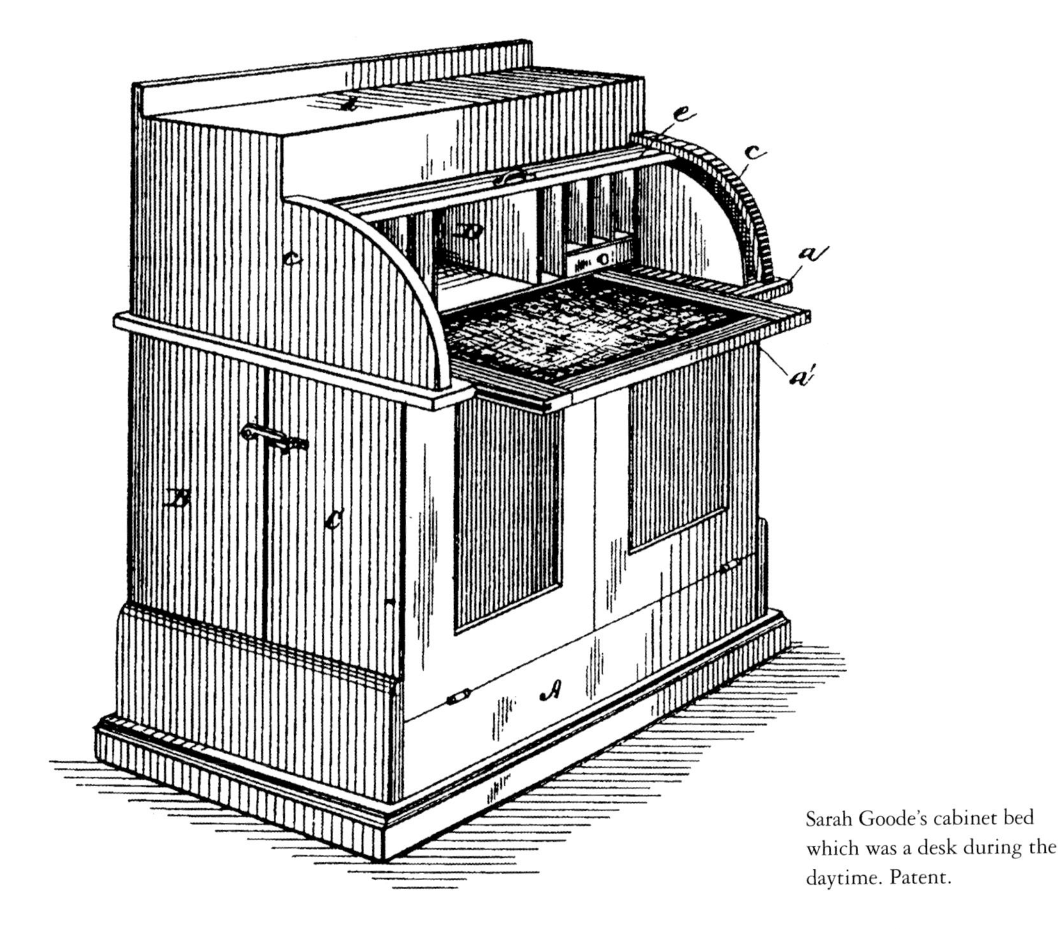

Sarah Goode's cabinet bed which was a desk during the daytime. Patent.

The multifunctional quality came into furniture design on a more practical level when Sarah Goode, the owner of a furniture store in Chicago, became the first Black American woman to hold a patent for her cabinet bed.[71] In the daytime this was a robust desk, and at night it ingeniously unfolded to make a bed. Another 'Dealer in Furniture', this time in Glasgow, Mary McAllister,[72] took out a provisional patent for a 'Combined Chiffonier, Cupboard or Chest of Drawers and Bedstead and Bed'. The chiffonier, cupboard and chest could all be folded down to create a bedstead and bed. Mary Claxton[73] designed a bed table and bookrest for use by invalids, which was exhibited at the Chicago World Fair in 1893. The table would rest across the bed and the bookrest could be pulled out at the front. Pockets hung down the sides to store extra material. Annie Keightley, a decorator from Middlesex[74] resolved the problem of limited space with her extending bed in which a single became a double.

The repeal of the window tax in 1851 meant that numerous, large windows could be included in the design of the new Victorian

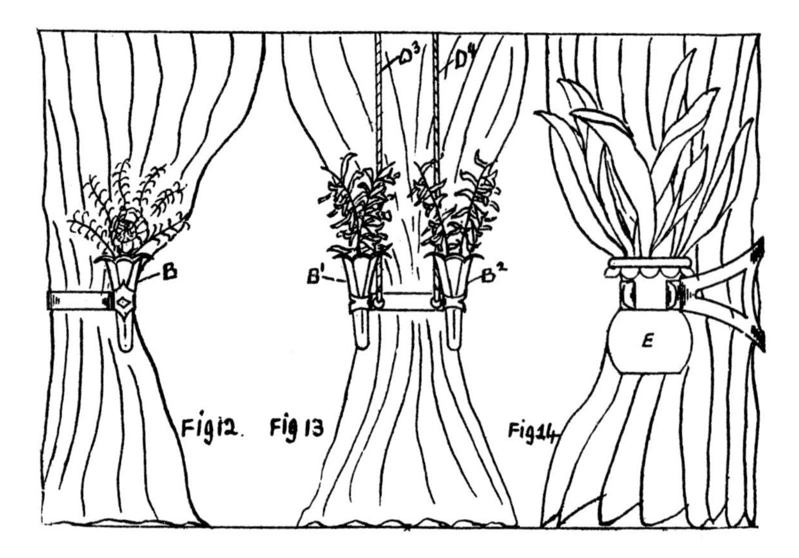

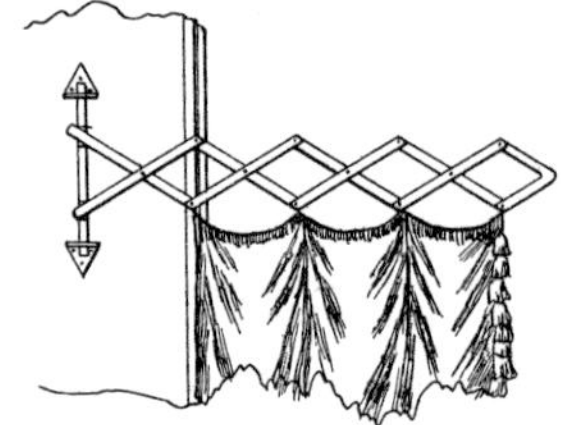

Hinged brackets from which Nelly Bacon hung curtains. Patent.

Plant pots and vases of flowers added extra decoration to Emma Stevens's curtains. Patent.

middle- and upper-class homes. However, those who chose the Gothic style, with its pointed arched windows, came across a problem to be resolved by the curtain and blind maker: that of how to cover them. Mrs Jemima Goode from the Isle of Wight had a noteworthy idea in 1843 which she submitted to the Society of Arts. She was, no doubt, one of the residents of the many new villas being built on the island, which became fashionable with the building of Osborne House, designed by Prince Albert in grand Italianate style. It was also the home of Alfred, Lord Tennyson, the poet laureate, and Julia Margaret Cameron, the photographer. Mrs Goode[75] was awarded a Society of Arts Mechanics Award for her 'Improved Gothic Window Blind'. This was a conventional blind for the lower, rectangular part of the window, with a separate one running from the top roller of the first blind, cut in the shape for the Gothic arch and pulled upwards into its apex.

Nelly Bacon, a 34-year-old journalist from London,[76] came up with an intriguing if cumbersome method to hang curtains, allowing for gradations of light to enter the room. The curtains were hung from hinged brackets at the side of the windows so that the 'swinging arms may be opened away from the window for the purpose of letting more light in or to enlarge the field of view, or may be swung right around and back on each side of the window'. The swinging arms could be hung in a zigzag manner to form angles, or fold vertically on each other. Emma Stevens from Rochdale[77] decided to make her drapery even more decorative by adding lacquered tin curtain ties into which dried plants could be inserted or even small vases and cut flowers.

By the turn of the century many people, including May Tennant from Bruton Street in London,[78] fully appreciated the benefits of ventilation in buildings. Her invention, in 1904, for 'Improvements in Air Filtering Ventilators,[79] combined an opening window and an air-filtering roller blind for both domestic and industrial purposes. Clean, dust-free air was essential:

> as the window is opened the space previously occupied by the window is automatically filled by a medium capable of acting as a filter to prevent the passage of dust, soot and the like while permitting the passage of air, and is especially adapted for use in towns in connection with the windows of laundries, workrooms etc and private houses where the exclusion of dust is of importance.

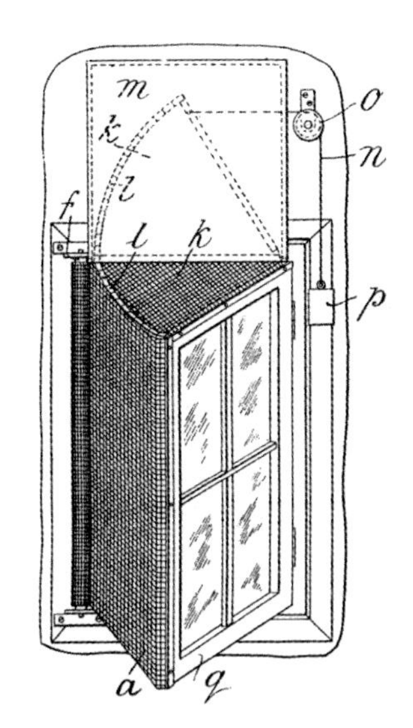

The window ventilation device invented by May Tennant. Patent.

The blind was made of gauze, muslin or wire gauze and its roller attached to the bottom of the window frame, the other end to the lower part of the sash window. When the window was opened the blind would be automatically unrolled to fill the open space. It could be adapted for fanlights and French windows.

Cleaning windows safely was the concern of Harriet Heron from Bexhill-on-Sea.[80] Her device consisted of a telescopic rod that could be locked into position to the required length. At the top end 'two right angled arms are secured, supporting vertically, a brush of sufficient size, while to the other side . . . opposite the back of the brush, a strip of wood or other suitable material is secured carrying a strip of rubber serving as a squeegee'. The window could be washed with the brush, dried with the squeegee that contained a strip of India rubber within its blade, and polished with a duster placed over the brush. It allowed the window to be cleaned without requiring the housemaid to be endangered by standing on the windowsill, and could also be used for cleaning shop windows.

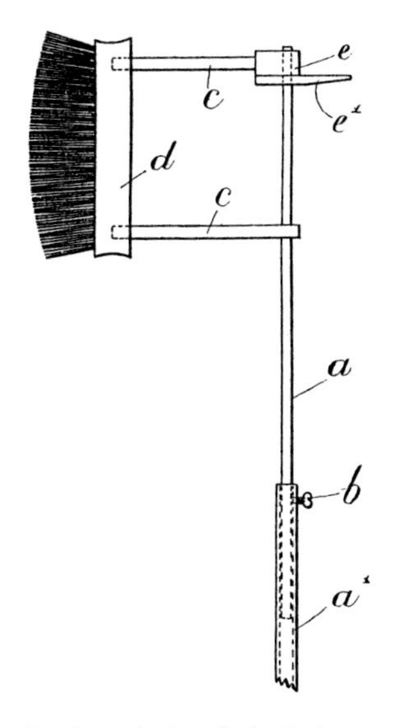

A telescopic-handled window-cleaning duster and brush by Harriet Heron. Patent.

Windows were regarded as a means of escape in the case of fire and Matilda Barron from East Molesey exhibited her rather dangerous fire escape at the Chicago Exhibition.[81] Having considered the problems of evacuation in case of fire carefully, she found

> The want of a ready means of exit from burning buildings has long been felt and many attempts have been made to devise successfully an efficacious apparatus for the purpose; but none have come into general use, on account of many requiring special structural contrivances, and others being costly and liable to get out of order and not workable at the critical moment.

Her wire or wicker dress stand was to be placed by the window at all times. When required as a means of escape the dress would be taken off it and the person climb in, a rope being attached to it and fastened to a strong ring on the floor. The basket and escapee were lowered out of the window.

The onset of the Arts and Crafts movement and women's admission to the new schools of art and design enabled them to set up studio industries and develop traditional craft skills, such as sewing, painting and weaving, into wider and more commercial fields. Many, like the daughters of the designer Christopher Dresser, were employed in their fathers' businesses and huge numbers found work in the pottery industries around Stoke-on-Trent. The Art School at Stoke was opened in 1847, offering classes for women and men to attend after a long day's work painting the latest motifs on the pots, tiles, fountains and bathroom ware.[82] In London, as a result of the success of the Great Exhibition and the vision of people like Prince Albert and Henry Cole, a school of industrial design opened in 1852. It was later renamed the Royal College of Art by Queen Victoria in 1896. Women's abilities to paint delicate and intricate designs and motifs on to ceramics were in advance of their male counterparts, and they quickly found work in the factories of Josiah Wedgwood and Herbert Minton. The scrumptious large dinner services, bathroom tiles and fountains for conservatories of the newly wealthy and impressionable middle classes were painted by thousands of working-class women. However, the design work was still regarded as men's prerogative.

MINTONS, LTD.,

China Works,

STOKE-UPON-TRENT, ENGLAND.

Potters by Special Appointment to Her Majesty.

CHINA, EARTHENWARE, MAJOLICA, PARIAN, CREAM-COLOUR, AND GLAZED TILES.

MESSRS. MINTONS have special Shapes and Designs adapted for the United States, which may be seen at all the principal China Stores.

Manufactory - - - STOKE-UPON-TRENT.

London Warehouse } 28, WALBROOK, E.C. (*opposite Cannon Street Station*).

Advertisement for Herbert Minton's factory from the *Official Catalogue of the Royal Commission for the Chicago Exhibition, 1893*. (*Author's Collection*)

Some women did attempt to break through these boundaries and set up their own studios but these were short-batch craft-based production lines. The potter Dora Lunn set up her own studio in West London after having trained at the Royal College of Art and worked for her father. The desire for independence enabled her to produce well-designed ranges of ceramics for the domestic market, very different from the somewhat opulent ware desired by the upwardly mobile middle classes. In 1893, when they married, Edith and Nelson Dawson combined their talents of her enamelwork with his silversmithing to produce a range of highly coloured, enamelled

Examples of the enamel and metalwork pieces designed by Edith and Nelson Dawson. (*Cheltenham Art Gallery and Museums/Bridgeman Art Library*)

metalwork which included jewellery, light fittings, bath taps, fire grates and wall plaques for the home, as well as church screens and crests and emblems for embassies and law courts. Their success was dependent upon each of their talents however; Edith received little recognition, even though she had published books on enamelling. It was her husband who was a member of the Art Workers Guild which denied women access and it was he who received more acknowledgement.[83]

Animals

The nineteenth century witnessed the growth of the zoological garden in most major cities. Exotic animals from far and wide were penned up and stared at by an enquiring public. In homes, too, pets became popular. Queen Victoria had her King Charles spaniels and Sir Edwin Landseer's paintings of wild and domestic animals adorned many walls. Horse and dog racing had long been popular, the course at Ascot having opened in the eighteenth century.

In Germany, Baroness Margarethe Johanne Christianne Marie von Heyden[84] and her husband were clearly concerned at weakening the

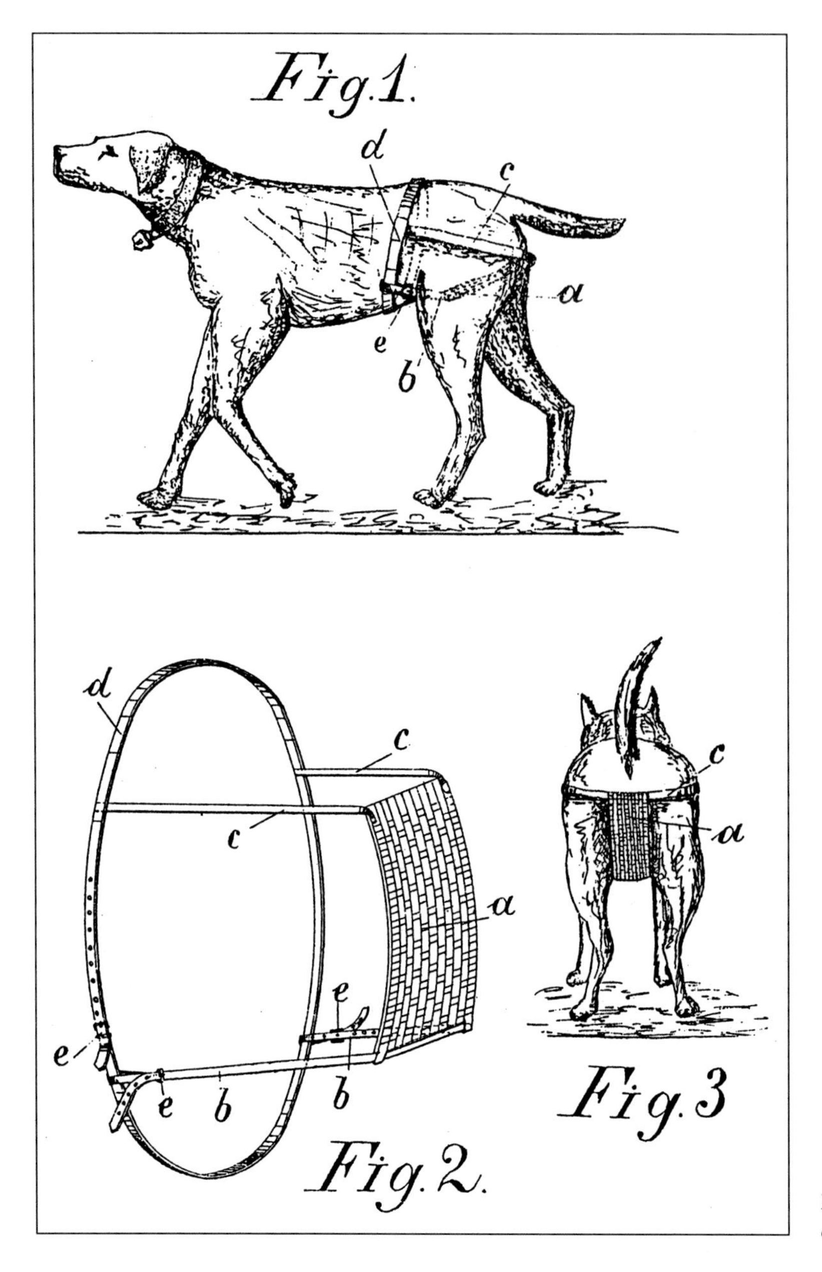

Margarethe von Heyden's canine chastity belt. Patent.

pedigree of their animals with cross breeding. They designed what could be described as a canine chastity belt to prevent 'coition in the case of bitches and other female animals more particularly for the purpose of preventing cross-breeding'. This was specifically for sporting animals and made of a shield of plaited material strapped to the animal's body to 'cover the genital parts without interfering with the animal's excretions'.

CHAPTER FIVE

'Amusement and Instruction': Designs for the Children

The notion of childhood as different from adulthood is modern. Until the twentieth century little attention was afforded to children. In an era before widespread use of contraception, families were large and many children died in infancy. The survivors were regarded as miniature adults and, according to social class, as inheritors of wealth, future wives or factory workers. Some children spent years separated from parents who had deserted them for exploits abroad as representatives of the British government or developing commerce overseas, described in detail by Katie Hickman in her fascinating book *Daughters of Britannia*.[1] There were also thousands of children living in abject poverty in Britain, eking out a living working in factories, mills, on farms and in domestic service and mainly illiterate until the government recognised the importance of a basic education. Others were orphaned or so poor they lived in the workhouse.

Lord Shaftesbury set up the 'Ragged Schools' in 1844 to provide education for the very poorest children. But it was not until the first Education Act of 1870, when board schools were established, that elementary schooling for the working classes was seriously considered. Children were then taught the basic skills as well as receiving a non-denominational religious education. Ten years later it became compulsory for all children between five and ten years of age to attend school. It was only in 1902, when local education authorities were introduced, that universal secondary education became a possibility. While the public schools had, for generations, educated the sons of the wealthy, the abilities of middle-class girls were at last recognised by people like Frances Mary Buss, who founded the North London Collegiate School in 1853. In 1871 Maria Grey, Mary Gurney and others formed the National Union for Improving the Education of Women of All Classes, not only to educate girls to a high standard, but also to provide employment for teachers. The union also aimed to help girls go into higher education and to change public opinion towards the realisation that they deserved as good an education as boys. The following year the Girls' Public Day School Trust established schools aimed precisely at these girls.[2]

On Sundays, in many households, games, toys and books had to have a religious basis; the Noah's Ark and stories with a strong moral element were popular. Many toys were expensive to buy and therefore only for the wealthy; until mass production, at the end of the nineteenth century, they were often handmade or miniature replicas made by craftspeople for their children. Dolls might be made from leather, pegs and rags; balls from straw plait and clay.

As early as 1801 the notion of an 'educational toy' was already being considered, albeit one for the very wealthy, by Ann Young, a cellist and writer living in Edinburgh. She devised her mahogany 'Box Containing Dice, Pins, Counter &c for Amusement and Instruction',[3] with its ivory dice, counters and markers, to enable children to understand the principles of musical notation. This was a

> new invented apparatus, consisting of an oblong square box, which, when opened, presents two faces or tables, and of various dice, pins, counters &c. contained within the same, by means of which six different games may be played, which besides being amusing and interesting, and such as children of eight years old may be taught to play, are at the same time an improving exercise upon, and serve to render familiar, and to impress upon the memory, the fundamental principles of the science of music, particularly all the keys or modulations, major and minor, both with common and uncommon signatures, musical intervals, cords, discords with their resolutions, and the most useful rules of thorough bass.

Within its lid a trade card states that it was dedicated to HRH Princess Charlotte of Wales and manufactured by Mair Woods & Co. of Edinburgh.[4]

Toy replica soldiers of the Crimean War were popular in the 1850s and Mary Ann Smith[5] from London devised a method to make these miniature figures, and dolls' house furniture, stronger by using bonnet wire covered with fine thread and silk, fastened to paste-board shapes. These were then covered in clay mixed with glue or size, which when carved looked like wood, or resembled cast metal when painted or varnished to add the finishing touches.

Obviously influenced by the skirmishes and battles of European empire builders of the period, a French spinster, Hermance Edan, introduced her 'New or Improved Game'[6] in 1909. Complete with a flag, four mines, one general commanding, one general of the

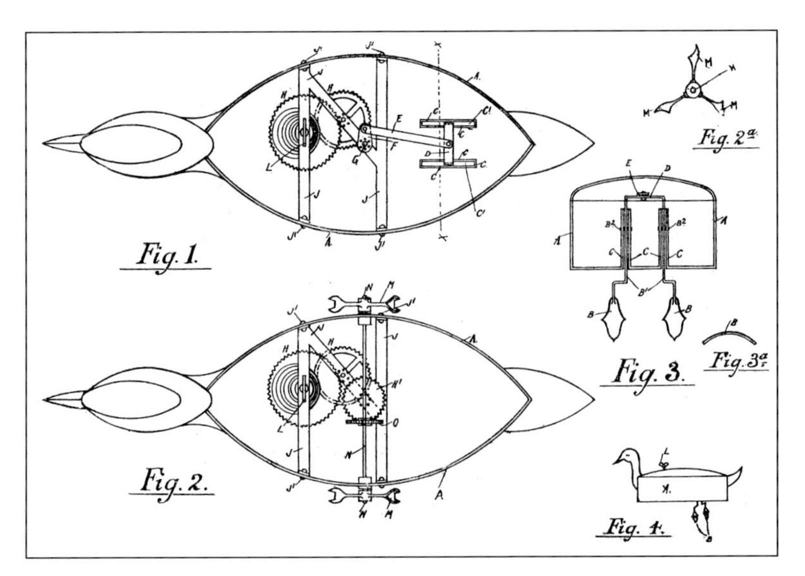

The automatic swimming duck by Laura Biddle. Patent.

brigade, two colonels, two majors, four lieutenants, four captains, four sergeants, four sappers, eight scouts and one spy, this game of strategy was 'a game for two persons . . . but the number of pieces, features, obstacles or conformations of the theatre of war may be unlimited, while the different marches can be varied infinitely. The winning of the game being the capture of the flag'. Mlle Edan's game was the basis of the internationally popular Stratego® game first marketed by Milton Bradley in the 1960s and played by both children and adults.

Automata had been popular since the mid-eighteenth century[7] when interest in the possibility of recreating a moving, human-like form caused consternation and philosophical debates on the meaning of life. Thomas Edison created a talking doll in 1890 and in 1897 Laura Biddle 'an automatic swimming duck'. The trick within any automaton was to hide the mechanism causing the movement. Laura Biddle hid it inside the duck's body, and its feet paddled to the delight of the audience.

Dolls, the creation of the human image, are universal in all cultures, whether as playthings, icons or sacrificial figures. The porcelain, bisque and wax dolls of the eighteenth and nineteenth centuries were highly sculpted and costly, making them available only to the wealthy. In poorer families, parents made dolls for their children carved out of wood or rags sewn together; children, too, made their own. One girl, in London in 1905, designed her own doll using an old black shoe, with paper features pasted on to its heel to make a face, limbs made of fabric-covered sticks, a sock for the hair, and scraps of material for the dress and apron.[4] This imaginative use of materials created a highly individual doll very different from those being mass-manufactured by doll makers, such as Käthe Kruse in Germany. In 1906 Käthe Kruse began to make muslin dolls for her own children and spent years perfecting a method to mass-produce them. This she finally achieved in 1911 with a method to cover the muslin with a special chemical formula and then mould it. The two halves were stitched together and stuffed with kapok to make a doll and the faces were painted individually. Not only did she sell them in Berlin but also hundreds were sold in the United States.[9] The beauty and simplicity of Käthe Kruse dolls are in sharp contrast to the Kewpie® doll. Kewpie® dolls, the idea of Rose O'Neill, originally appeared in 1909 as fun drawings for adults which were published in the American *Ladies' Home Journal*. Rose O'Neill saw the potential to transpose them into rag dolls for children and adults. In 1912 she had registered the design[10] and the dolls were mass-produced firstly out of bisque and later celluloid. The Kewpies®, with their large ogling eyes, smiles, pointed hair and tiny wings became a craze and millions were sold in the USA and Britain.[11]

Beatrix Potter and William Heelis on their wedding day, October 1913. (*John Heelis*)

The children's author, Beatrix Potter, brought a refreshing repeal of the often religious and moralistic undertones of Victorian children's literature with her books, which are still widely read by children throughout the world. Her unsentimental approach to anthropomorphism, clear watercolour illustrations and the appeal of the naughty Peter Rabbit escaping from the cabbage patch and the hands of Mr Macgregor, continue to transfix children. Beatrix Potter became one of the first people fully to exploit character merchandising when, in 1902, she gave her publisher, Frederick Warne & Co., permission to register the design for Peter Rabbit as a soft toy. The now battered sepia photograph of the original Peter Rabbit, in the Public Record Office,[12] reveals a friendly if buxom rabbit, complete with waistcoat and leather slippers, with droopy ears and a slight smile on his face. In 1904 a Peter Rabbit board game was devised and the Beatrix Potter characters continue to be

Margarete Steiff in her factory. (*Margarete Steiff GmbH*)

nurtured by generations of children throughout the world; their tales and the illustrations are timeless.

Around the same time, in the United States, President Theodore (known as Teddy) Roosevelt went bear hunting in Mississippi. The only bear available to shoot had been confined to a small space and fastened to a tree. When the president refused to shoot it, cartoons appeared showing his dilemma. Although bears had previously been produced as automata and soft toys, they had not, until President Roosevelt's encounter, acquired a name; from then on the mass appeal of the 'teddy' bear began. One woman with a long and close association with the teddy bear is the German, Margarete Steiff. Disabled by childhood polio, she was a very capable seamstress and in 1880 opened a mail order company selling soft toys. Her soft bear had provoked little interest until she exhibited it at the Leipzig Fair in 1903 and Herman Berg, a toy buyer from New York, decided to buy 3,000 to satisfy American customers eager for teddy bears. So successful was the American market that the company called the period from 1903 to 1908 the 'Barenjahre' (bear years) when its annual production rose from 12,000 to 975,000.[13] Maybe it is the simplicity of the characterisation of the Steiff animals that has kept many of them in production for years and, like Peter Rabbit, enthralling generations of children.

Margarete Steiff with one of her first bears. (*Margarete Steiff GmbH*)

The kindergarten movement slowly developed from the mid-nineteenth century in Europe, when Johann Pestalozzi in Switzerland and Fredrick Froebel in Germany began to promote the sensory needs of the under-fives. As early as 1816, Robert Owen, believing in the benefits to young children of a stimulating environment and fresh air, opened a nursery for the children of the workers in his cotton mill in Scotland. In 1880, Mary Tyler Foote from Boston, Massachusetts patented her 'Game Apparatus'.[14] She states that this was an educational game to teach multiplication and contained thirty-six blocks in rows of three.

Finally, one woman whose name has dominated twentieth-century pre-school education is Maria Montessori. Her appreciation of the psychological and spiritual development of children, in relation to their physical well-being, was quite revolutionary, and very different from the British approach of 'children must be seen and not heard'. She was an Italian doctor, whose work in a children's psychiatric clinic gave her a basis from which to apply, to all children, her methods of carefully planned and stimulating sensory development. In her handbook, published in 1914,[15] she acknowledges the enormous benefit that the new hygiene practices had on children's physical health, but notes the lack of appreciation of their psychological and emotional development. The newborn child was equated to an immigrant in a new country, unable to speak the language or understand the strange customs. Her plan was to introduce her or him to the world gently, and at the child's pace. The children should be together in a special place, preferably a 'Children's House' with a series of rooms and a garden to get fresh air; all furniture and objects were to be on a child's scale. Central to the house was a room for 'intellectual work' where her specially designed equipment was used.

Dr Maria Montessori looking at a new edition of *Il Metedo della Paedagogia Scientifica* in 1913. (© *Association Montessori Internationale*)

While some of the equipment might appear rather rigid by today's standards, creativity and expressionism were encouraged.

> Round the walls of the room are fixed blackboards at a low level, so that the children can write or draw on them, and

> pleasing . . . artistic pictures, which are changed from time to time as circumstances direct . . . a large part of the floor must be free for the children to spread their rugs and work upon them . . . statuettes, artistic vases or framed photographs, should adorn the walls; and above all each child should have a little flower-pot, in which he may sow the seed of an indoor plant.[16]

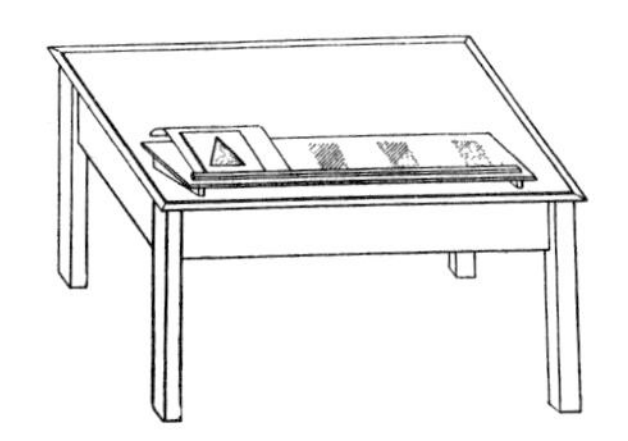

The sloping reading table advocated by Maria Montessori. Patent.

Maria Montessori was awarded three patents for her teaching equipment. One, in 1912[17] for apparatus to teach children to write consisted of a sloping reading desk to stand on a table and metallic plates with cut-out geometric shapes on it. Children of four years would be attracted by these brightly coloured shapes and by tracing them they would quickly become proficient in holding a pencil and writing confidently. This was further developed in 1914 with more complicated polygons designed to teach geometry.[18] Her equipment to teach arithmetic[19] consisted of an abacus with coloured beads and sheets of lined paper on which to write the results. Apparatus to teach multiplication and the powers of numbers was also attached.

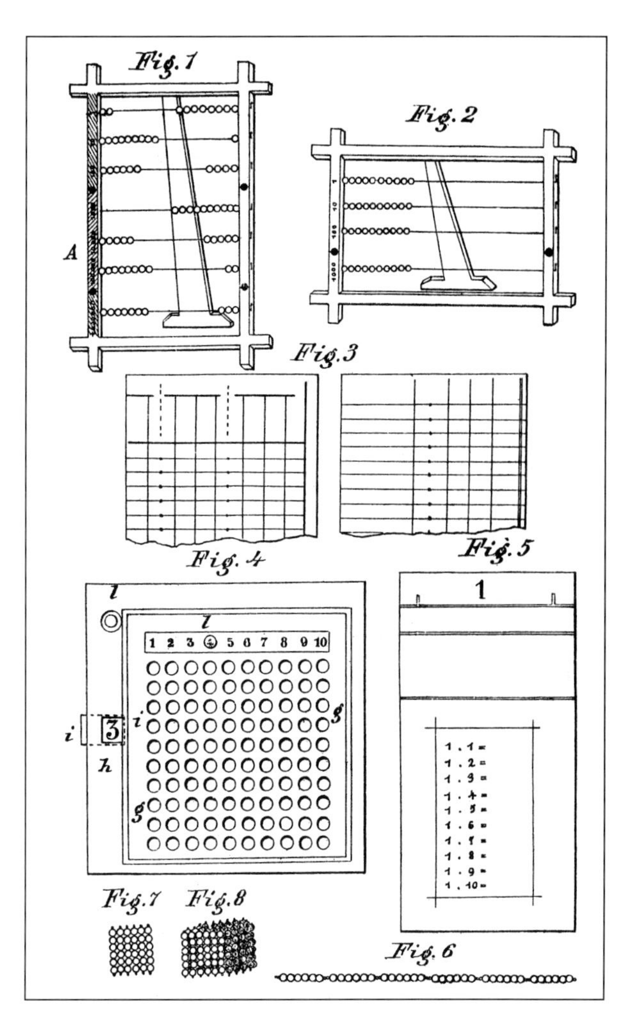

Maria Montessori's equipment to teach basic numeracy. Patent.

Although Maria Montessori is probably the most famous and influential advocate of the kindergarten movement, there were other women who designed intriguing educational aids for their schools. Augusta Pattersson, a teacher from Essex, designed 'An Apparatus for the Thorough Teaching of Numbers from one to twelve inclusive'[20] and Margaret Kay, an elementary school teacher from Wandsworth, a frame to present the basics of arithmetical notation for nursery and infant school children.[21] Emily Wynne-Jones, a certified teacher from Ilford, designed a 'Kinder-garten Tree'[22] made of wire and covered in loops and hooks. It could be used to hang drawings for nature study or for arithmetic when counters could be attached. Children could also decorate the tree as they wished. Taking an appreciation of nature even further was Florence Zambra with her toy garden.[23] This fantasy garden was made of a base board covered in a green fabric such as velvet or plush; sand or gravel were glued to it to make paths; small artificial flowers 'planted' and arches and arbours could be constructed.

The changing attitude towards children also impacted in other areas and towards the end of the nineteenth century a more relaxed approach, as with adults, was adopted in the design of children's clothing. The custom had been for centuries to swaddle babies; and stays and corseting for children were not uncommon in the eighteenth and nineteenth centuries. Roxey Caplin had designed corsets specifically for children, which, like the deportment board, were intended to keep the spine straight. Agnes Drury, who described herself as an authoress and inventor, was born in Madras in 1852, and lived later in Bath and London, where she devised her fascinating series of clothes in 1901.[24] Central to her clothing designs was the wrap-around style, so that babies did not have to suffer the discomfort of having tight clothes pulled over their heads, and could be dressed easily. Any fastenings were at the sides and ties were used as far as possible. The clothes could be made larger or smaller, used for

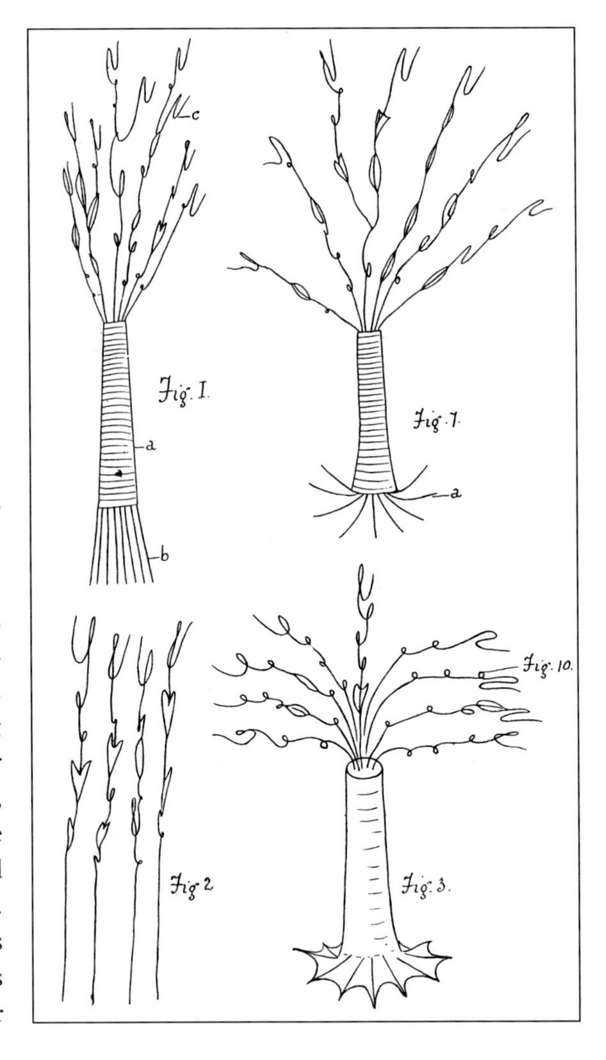

Emily Wynne-Jones's wire tree for use in kindergartens. Patent.

Fantasy miniature gardens were designed by Florence Zambra for children to explore. Patent.

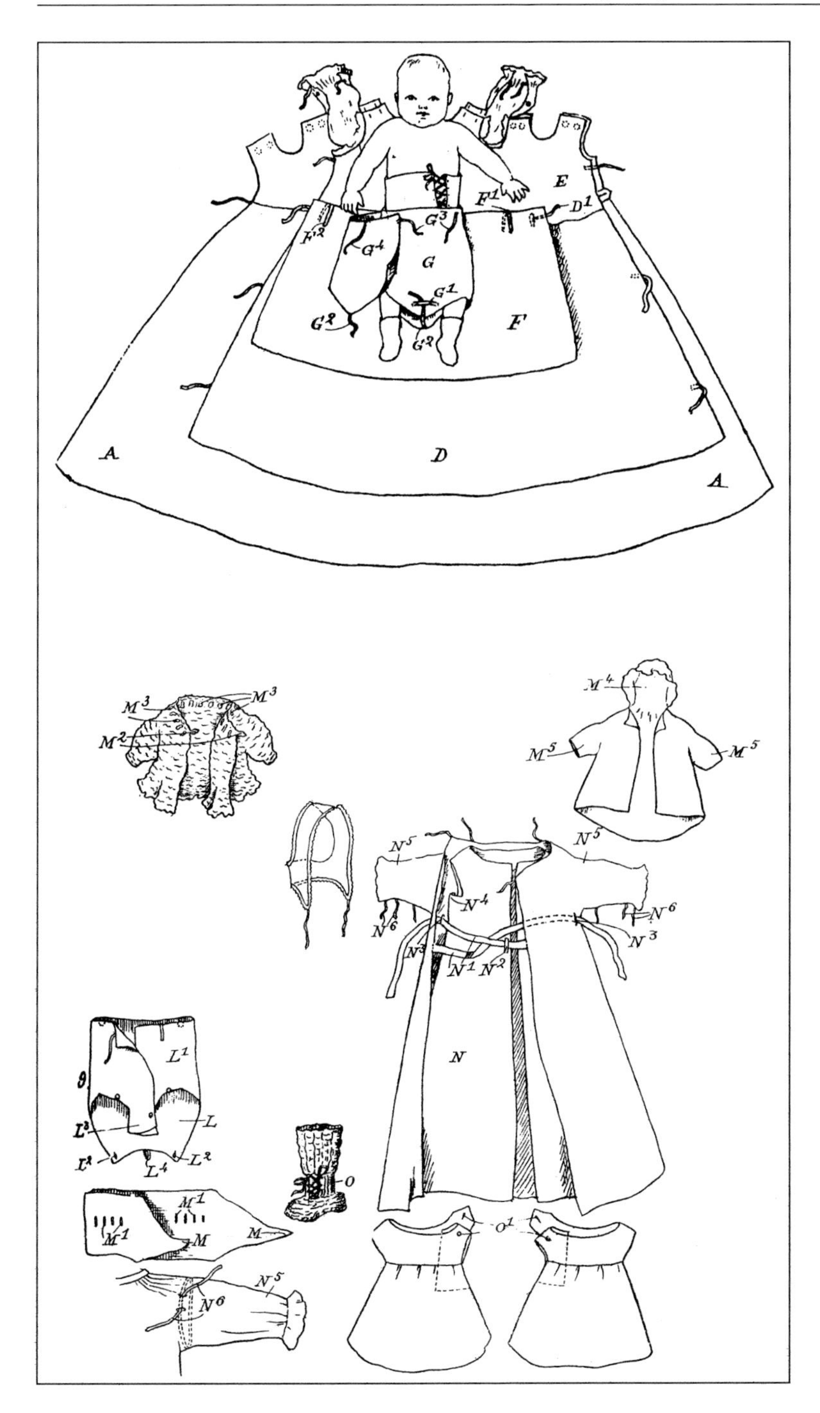

Part of Agnes Drury's extensive, wrap-around, clothing scheme for young children. Patent.

day or night wear and adapted to suit the size and specific requirements: sleeves unbuttoned and were removed from the shoulder, skirts could be lengthened or shortened. She even included skirts for women and trousers for men in her twenty-eight diagrams to show the variety of her capsule wardrobe scheme.

Magdalena Laue and inventors from Halle in Germany introduced a baby's nappy, made from one piece of India rubber or American cloth with a tie fastener.[25] The prime intentions were to prevent bed-wetting and the baby becoming uncovered when kicking: a very different approach from the centuries of swaddling which had prohibited any movement of active young limbs. She found that the cumbersome rectangular or square napkins did not fasten properly, were difficult to take off and often resulted in a wet bed.

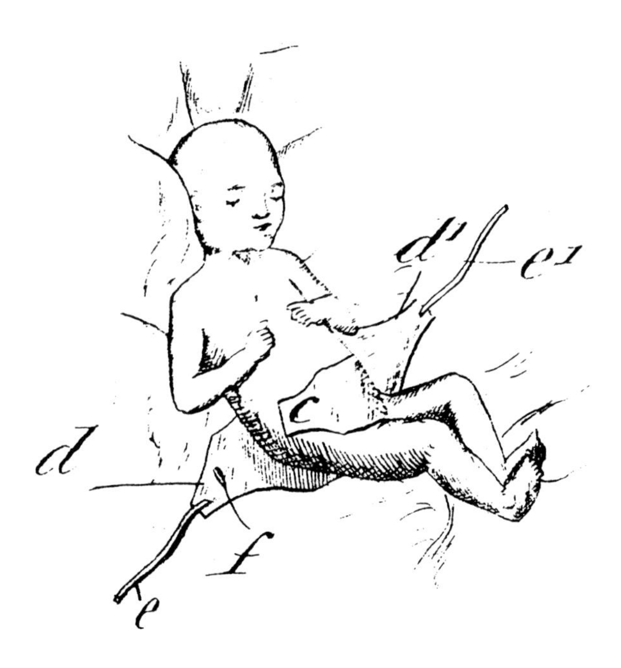

The India rubber baby's nappy invented by Magdalena Laue. Patent.

Keeping a toddler occupied in safety is a challenge for all parents and nannies. Ann Littlejohns Panter from Grays Inn Road in London devised an 'Improved Seat for Babies or Young Children'[26] in which a cushion was placed on a circular base board mounted on castors, so that the child could be pushed around easily. A cushion with a back, like a small armchair, was fastened to it and the child sat in it with a strap between its legs fastened to the base. A tray could be attached to the board, in front of the child, and toys placed on it. In Australia, Muriel Binney from Elizabeth Bay in Sydney devised a 'Portable Safety Playground for Infants'[27] and, based on the same principle, a folding cot.[28] Safety was an uppermost consideration in their design. The playpen was made of a collapsible frame of metal or wood, for indoor or outdoor use by one or two children, 'to prevent them from straying or injuring themselves, or being interfered with by animals'. The sides were covered in netting so that if a child fell against it, it would act like elastic and prevent the child from falling through the gaps. All care was taken to ensure that joints and hinges were smooth and the child would not be injured. A mattress was placed inside the frame to make a cot. When folded up, it occupied little space and could be easily carried.

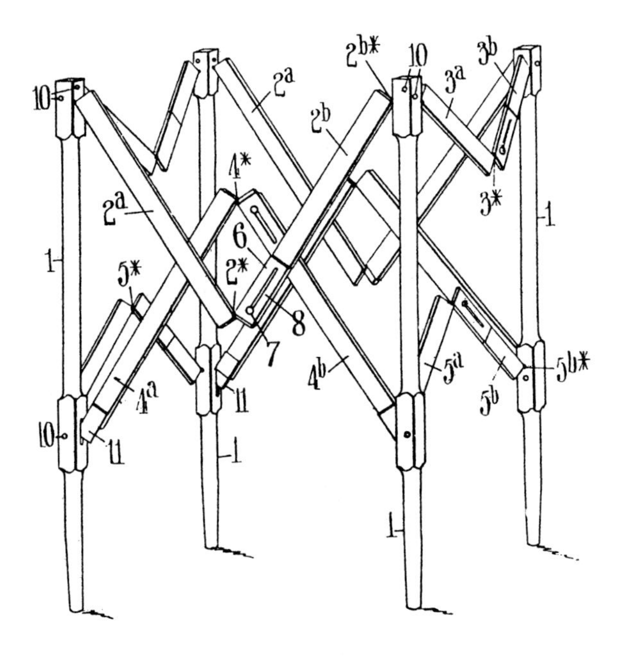

Part of the folding, safety playpen from Muriel Binney. Patent.

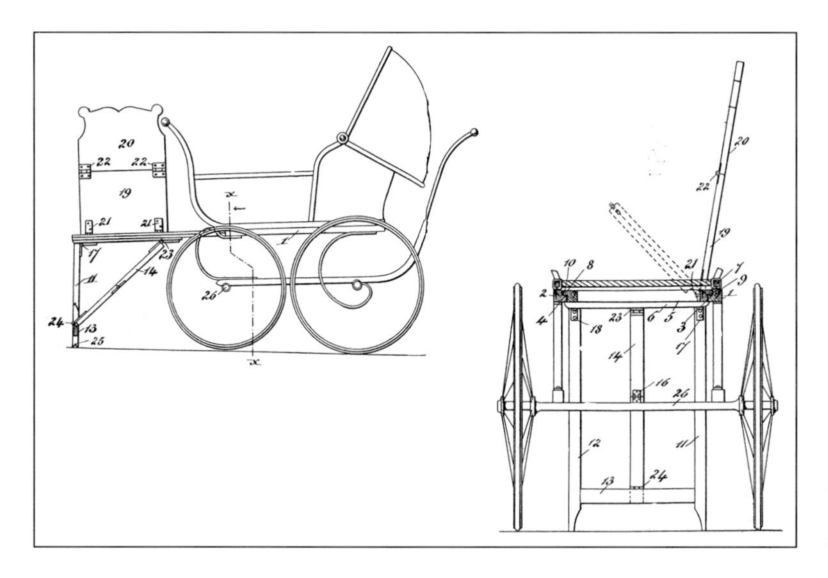

Anna Schaefer added a folding seat and picnic table to a baby carriage. Patent.

The first mass-produced perambulators, or 'prams', began to appear in the 1850s and, by the end of the century, the conventional four-wheeled, high-sided baby carriage was common. However, Anna Schaefer of Independence, Missouri in the United States had a novel idea for an accessory.[29] This was to add a folding seat and table to a pram so that 'a nurse or other attendant may be seated when desired and, subsequently fold the seat for concealment beneath the carriage', If partially folded, it could be used as a carrier and adapted as a picnic table.

One of the saddest items to be found in this search is the funeral carriage and hearses designed by an artist from Sydenham, Mrs Louisa Brown, in 1907.[30] These were specifically for the funerals of children in which the child's coffin was placed in an open or glazed receptacle on the roof of the horse-drawn carriage. Numerous hooks were attached to the coffin holder upon which flowers, wreaths and other decorations could be hung. Despite a better understanding of the spread of disease, since Chadwick's 1842 report, the infant mortality rate was still high as epidemics of scarlet fever, diphtheria, cholera and tuberculosis decimated families. In 1911 alone, 32,000 babies died in Britain from 'summer diarrhoea'.[31] Protection from the likely risk of infection, not only among the mourning family in the carriage, but also from the corpse, is clearly the main feature of Mrs Brown's proposals. She states that in previous hearse and carriage arrangements the coffin was placed under the driver's seat, which she thought not only 'insanitary but irreverent'. By placing it on top of the carriage, not only did the deceased child receive the respect it deserved, but the 'evils' of cross-infection were hopefully, avoided.

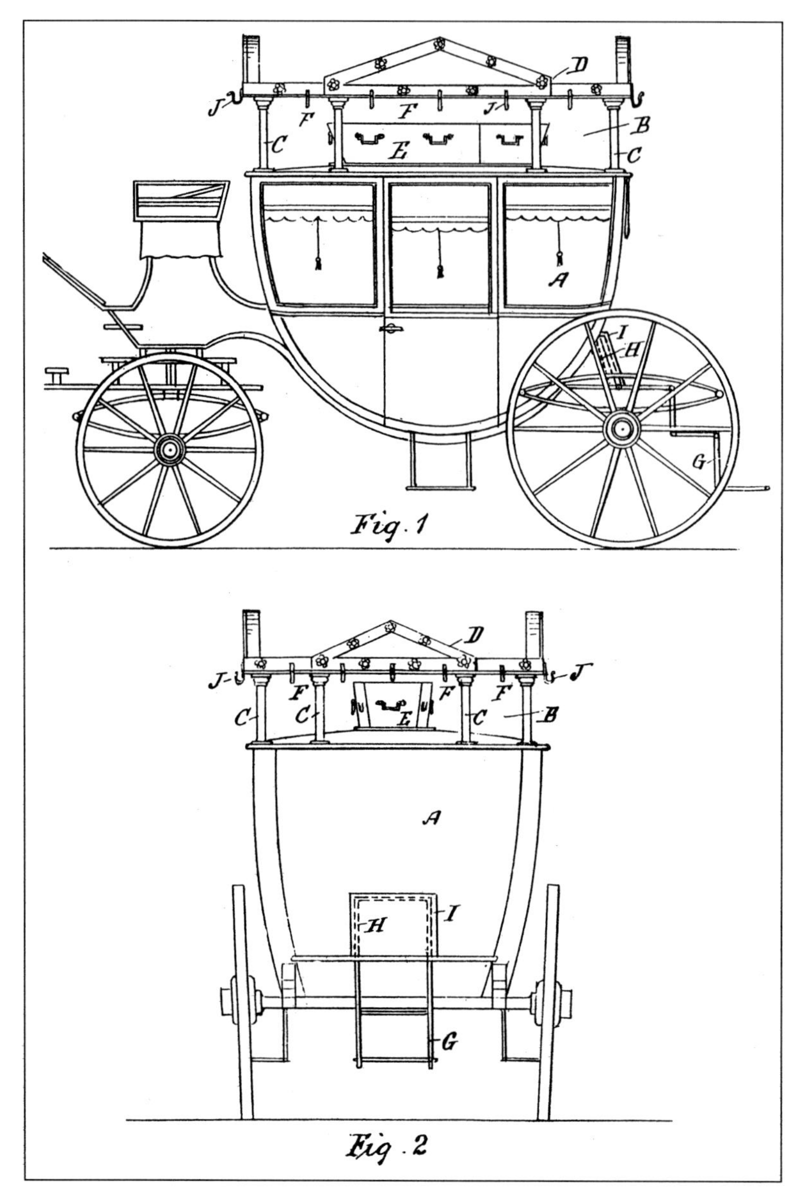

The hearse for children in which the coffin is placed, with dignity, on the roof, and hygienic fittings in the carriage ensure that the passengers will not become infected with disease. Patent.

Sanitary arrangements were also considered in the interior design of the carriage which she suggested should be painted in enamel rather than upholstered, with 'detachable seats and back cushions provided, covered with washable material; so that the interior of the said carriage may be thoroughly washed and disinfected after use'.

Using their direct experience as mothers, teachers and doctors, women made very practical contributions to children's lives during the period of transition from the miniature adult, with few rights, to the modern concept of childhood.

CHAPTER SIX

For the Greater Benefit: Philanthropy and Reform

The desire to fulfil a philanthropic role, to redress society's injustices and provide for the needs of the poor, galvanised numerous upper-class women with a social conscience into action during the nineteenth century. The Quaker Elizabeth Fry was appalled at the conditions endured by women incarcerated in Newgate Prison when she first visited in 1817, as well as the plight of those being sent as convicts to Australia, the majority of whom were in custody for petty crimes. Elizabeth Fry devoted her life's work to prison reform. Baroness Angela Burdett-Coutts, granddaughter of the banker Thomas Coutts and a friend of Charles Dickens, drew on her wealth to fund educational establishments and housing for the poor. John Ruskin funded Octavia Hill's first housing development for the poor in Marylebone, London in 1853. A great believer in the necessity for open space, Octavia Hill trained others to run similar projects and in 1895 became one of the founders of the National Trust for Places of Historic Interest or National Beauty, which remains one of the foremost organisations in preserving Britain's natural and architectural heritage. But it was Emmeline Pankhurst who would champion the cause of all women, that of suffrage. For forty years she was the figurehead, assisted by thousands of other women from all parts of society and the country, determined to change society's and government's attitudes towards women with their demand for the right to vote. Her husband, also a reformer and a friend of John Stuart Mill, was the author of the Married Woman's Property Acts in 1870 and 1882. The suffragettes suspended their activities during the First World War but resumed them immediately afterwards and eventually women received voting parity with men in 1928, a few months before Mrs Pankhurst's death.

Some women did find ways to combine the philanthropic and suffrage causes with the new technologies in their attempt to improve the lives of the poor. Notable among them was an authoress, editor, publisher and designer, Amelia Louisa Freund. Amelia Freund has left a fascinating record of her activities but regrettably receives barely a mention in women's history. She lived in London, traded as Mrs Amelia Lewis and in the 1870s edited and published a weekly newspaper, *Woman's Opinion*. She wrote a series of twelve booklets called *Food Leaves*, patented three different cooking ranges and manufactured them as well as planning another publication, *The National Food and Fuel Reformer*. Clearly influenced by the suffrage movement, *Woman's Opinion*, which only lasted for a few months, contained numerous articles for women. Not only are the products of Amelia Lewis, Mr Rimmel's 'New Perfumes' and Rowland's 'Macassar Oil' advertised, but there are also illustrative articles, mainly penned by Amelia, which at times read like sermons to perfect living. Probably influenced by Octavia Hill, she wrote an article demanding an improved playground for children in Leicester Square, citing the need for open space and the detrimental effects of urban living. She called for Parliament to demand that the Metropolitan Board of Works turn

Portrait of Angela Burdett-Coutts from *Vanity Fair*, colour lithograph, nineteenth-century English School. (*Wimbledon Society Museum of Local History/ Bridgeman Art Library*)

> the unsightly waste into a humanising sight . . . It is most a cruelty to let little ones be born, and brought up in an atmosphere of those dreadful places – modern large towns – and cramp the nascent strength and growing energies of children, into taking a straight walk

Left: Spitalfields soup kitchen in 1867. (*Illustrated London News*)

Opposite: A suffragette on hunger strike being force-fed by prison officers in 1912. (*Illustrated London News*)

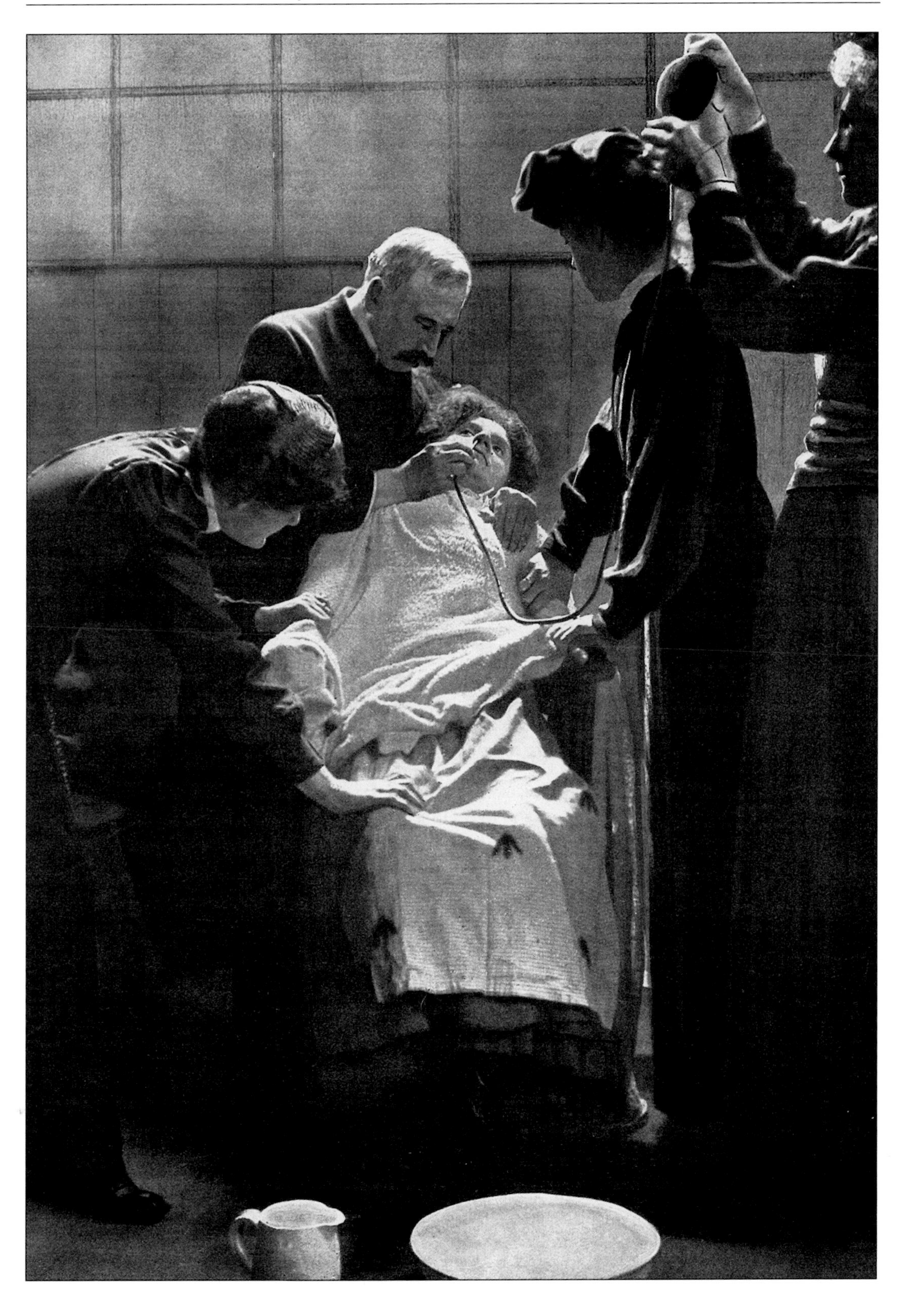

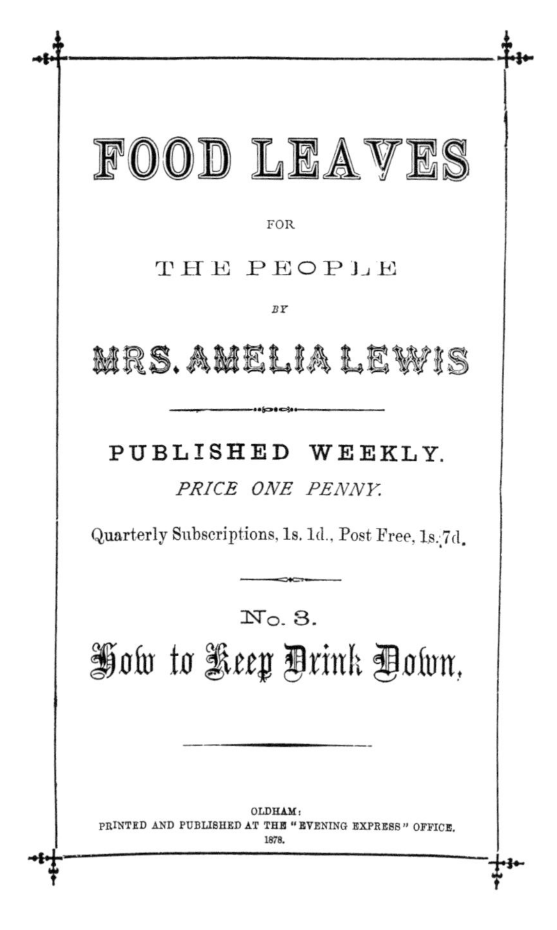

FOOD LEAVES

FOR

THE PEOPLE

BY

MRS. AMELIA LEWIS

PUBLISHED WEEKLY.

PRICE ONE PENNY.

Quarterly Subscriptions, 1s. 1d., Post Free, 1s. 7d.

No. 3.

How to Keep Drink Down.

OLDHAM:
PRINTED AND PUBLISHED AT THE "EVENING EXPRESS" OFFICE.
1878.

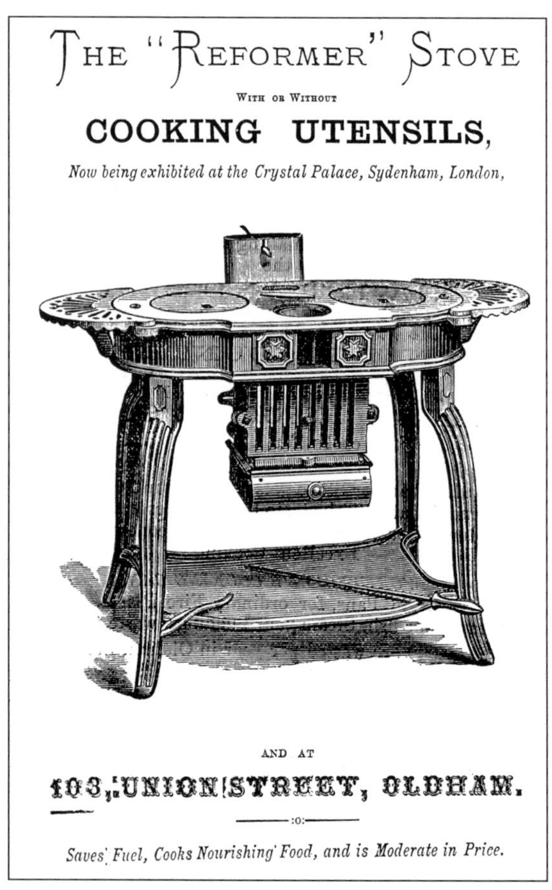

Above, left: Title page of one of Amelia Lewis's *Food Leaves*: *How to Keep Drink Down*. (*Oldham Local Studies and Archives*)

Above: An Advertisement for the patented 'Reformer Stove'. (*Oldham Local Studies and Archives*)

> up one street and down the other . . . into abstaining from running after each other.[1]

It was, however, her fundamental belief that nourishing food cooked in an economical way would improve the lot of the poor, which underpins all her work. Clearly influenced by the Temperance movement, she fiercely denounced the effects of drinking alcohol in *Food Leaves*.

Not satisfied with her publishing and writing career, Amelia Lewis became a designer and manufacturer of cooking stoves, for the poor and the army. Two – the 'People's Stove'[2] and the 'Reformer Stove'[3] – were for domestic use, fuelled with peat, enabling them to be free standing and not dependent upon a fireplace or gas for power. Like many Victorian inventions they had a dual purpose: to cook food and also to heat the homes of the poor. 'The People's Stove' came in three sizes, the smallest costing around 12*s*. Her goal, to alter the eating and heating habits of a whole class of people, was a formidable one. As a commentator from the contemporary magazine *IRON* reported, albeit in Mrs Lewis's own journal *Woman's Opinion*,

> The mere introduction of economising fire-places is, however, but a small part of the task set for herself by Mrs Lewis, who aims at reforming the entire cooking arrangements of the lower classes, by means of the instruction of the young, lectures to the old, and the publication of a journal devoted to the question of 'food reform'.[4]

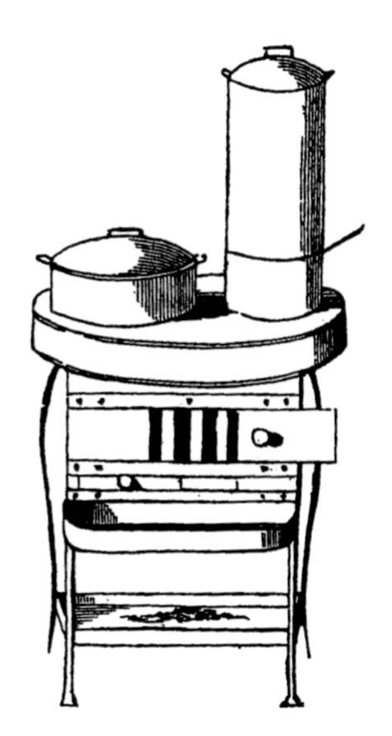

'People's Stove' Patent.

She launched the 'People's Stove' at a meeting in Southampton Street, off The Strand, in London in March 1874 by giving a cookery demonstration: 'the crucial culinary test – the cooking of vegetables – was undergone most successfully.'[5] The audience was also impressed by the stove's ability to heat a room. Steaming food in a suitable pan or using dry heat directly from the stove was her preferred way to cook in order that 'food (be prepared) for human consumption so that the human stomach may be able to digest the same, or assimilate the substances contained therein to the requirements of the body more easily than if the food were prepared according to the system or methods hitherto in use'.[6] By making the peat fire as compact as possible she endeavoured to retain its heat and disperse it evenly across the hob and into the cooking pots. Her second stove was more elaborate but on the same principle as the 'People's Stove'. The 'Reformer Stove' was exhibited in one of the many exhibitions held in the Crystal Palace, Joseph Paxton's glass

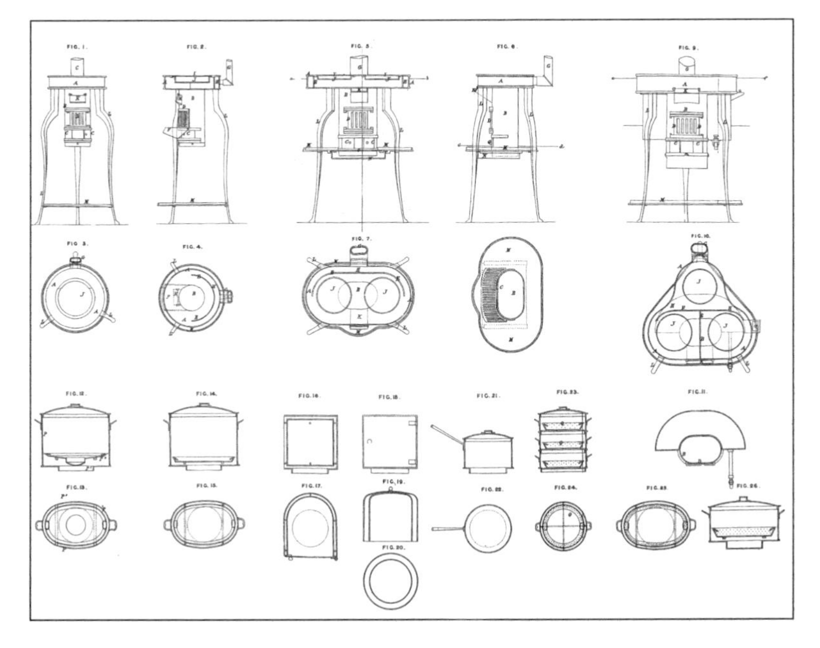

Patent diagrams for Amelia Lewis's 'Reformer Stove'.

exhibition hall that had been relocated to the new suburb of Sydenham in South London.

Four years later, in 1878, Amelia Lewis decided to tackle the problem of feeding large numbers of soldiers in the field and designed a stove for military use.[7] Here she again used the principle of the portable stove, standing it on legs. This stove contained a chamber for the fire with a sliding, perforated door, to regulate the draft, while the smoke was extracted through a flue. All the stoves were made at 103 Union Street, Oldham, a town which at that time had many stove, grate and kitchen range manufacturers. It is probable that Amelia Lewis lived in London, her publishing address was in Southampton Street, and she used the Oldham address as a manufacturing and depot base for her business.[8]

Amelia Lewis's series of booklets, *The People's Food Leaves* comprised about sixteen pages and cost 1*s* 1*d* a quarter. Their titles included: *The Purpose of Food*, *How to Keep Drink Down*, *The Waste of Food and What Wages can Buy*, *The Cooking of Food*, *How Children Should Be Fed* and *How to Save Fuel*. In them she expounds, rather like a preacher, her philosophy of cooking for healthy living. Where Roxey Caplin had demanded good posture and respect for the body to maintain health, Amelia Lewis's mantra was well-cooked food:

> Food plays so important a part in our life, that if we do not provide for it properly, we may lay the root of disease, ill health, and loss of strength. We have the two things before us. A certain amount of means to do with, and the proper spending of these means. I scarcely believe that it is possible for those in high and middle class life to understand what it is for a workman's family to live off £2 per week . . . or even on less that is . . . 12*s* per week.

She continues to describe what happens if the body is deprived of fuel:

> We lose substance, because our body is like a machine using up power; this power must come from somewhere, just as it comes from the steam engine . . . or power is drawn from the blood.[9]

and goes on to describe the power of digestion to feed the blood.

As strongly as she felt about food, it was, however, the effects of alcohol which aroused most passion and vitriol in Amelia Lewis. The drunken husband arriving home, having spent his wages, abusing his wife and children has been the subject of dramatic scenes in numerous novels. The wife in turn would have to make ends meet with less money than the paltry sum her husband had earned.

Although the Matrimonial Causes Act of 1857 had finally enabled women to divorce and keep some contact with their children, and part of their property, it was not until the Married Woman's Property Acts that they could begin to contemplate financial independence. Had she left an abusive relationship, the workhouse was probably the only door open to a woman and her children. Small wonder that the hat-making women of Luton treasured their financial independence.

In the eighth *Food Leaves* entitled, *How to Keep Drink Down*, Mrs Lewis describes the detrimental effects of alcohol to the body's functions. Like many at her time she regarded the wife and mother to be responsible for her family's well-being, especially in the area of alcohol. She claimed there was a relationship between giving tea to young children and the future alcoholic, and that by using her meal plans and recipes, as well as support from a higher level, drunken husbands would be weaned from the demon drink:

> The educational department of the Queen's Government, the school boards . . . the church . . . the colleges . . . the magistrates . . . all should labour in this great cause, and help to make people understand what drink means, and why it has become the curse of our life . . . It is fearful, a horrible ghost this drunkenness, this constant swallowing of things, which we really do not want . . .
>
> . . . every mother who gives her infant, her child or her son and daughter tea for breakfast constantly, is making a possible future drunkard, for she makes the child's blood poor, she gives the child no true nourishment in the morning: but on the contrary, introduces a substance that weakens the child. Let this go on for years, let this child have to grow up and begin the hard work of life, and the want of strength in his or her system may lead him or her to fly to that quick means of getting false strength, which ends at last in drunkenness.

She continues, now in a religious tone, predicting disaster if the 'Creator' is abused by the demon of alcohol:

> The more we learn, the more hideous will appear to us the crime to waste powers given to us by the Creator himself to lower ourselves below any of the brutes. The brute has instinct and following that instinct eats and drinks what necessity requires, man to whom the use of will has been given [and] abuses that will and with it actually follows out a course, that must be destruction to him.

As in much of her writings, having outlined the problems in graphic detail, Mrs Lewis makes recommendations based on cooking appropriate food. To overcome alcoholism she states:

> To keep drink down in any family, you must keep food, and good food up; you must positively make up your mind to spend your money on food and not on drink alone . . . A woman complained . . . [in] Oldham . . . that I did not tell her how to keep her husband sober . . . she expected me to give some universal receipt. There are none, although I believe, that I will in time be able to give decoctions of herbs, that will have the effect, to create distaste for alcohol. The best receipt I could give her was to study her husband's food. To make it savoury, easily digestible, supplying sufficient nourishment or nitrogen and staying him almost without his knowing.[10]

According to Amelia Lewis, the tartness emitted from spicy food led to a want for alcohol, no doubt to quench the thirst. Equally she found that bland, flavourless cooking 'excites the workman to seek drink' probably in a search for stronger flavour. The wife's duty was to provide just the right food to eradicate the wish for strong drink. This meant good ingredients and appropriate cooking; undercooked or overcooked foods led to the craving; care had to be taken in roasting, frying and grilling so as to preserve the flavour. A soup, simmered for a short time to draw the flavours from the meat, with barley and vegetables added, would be ideal and full of flavour and satisfying to the palate, quenching any thirst for alcohol. This refinement of taste that had occurred among the upper classes and the French for generations had eluded the working classes in Britain. Her description of the effects of poorly cooked food reveal a catalogue of unsavoury and unappetising meals. Little wonder then that the desire for alcohol was evoked not only for its ability to alter the mental state but also the taste:

> A good stew is a most delicious dish, but when almost done to death it becomes tasteless, if not nauseous to many, and they turn from a half made meal to get some sudden strength out of beer. Underdone food has the same effect; half raw meat has no flavour. For the flavour comes through the thorough heating process; a woody cabbage lies very heavy in the stomach; badly baked cloggy bread cannot be dissolved but with great trouble, and the tired organs tear away till their possessor is driven to seek comfort in drink.

Mrs Lewis was heartened by reports of groups of young working men enjoying her soups and bread in an eating-house in the Oldham district of Mumps and believed that they would not succumb to the drink if they followed her regime. She recognised the attraction to alcohol of an increasing number of exhausted women and their belief that it boosted their sapped energy. Like Roxey Caplin she advocated physical exercise and her own good nourishing food were to be the therapies. At the end of the booklet, almost as a reward, she suggests recipes for these women: steak and onion stew, roasted rolled beef, boiled potatoes and cabbage, lemon suet pudding with sweet sauce. All had to be cooked with just the right amount of heat, which, of course, could be supplied by one of her stoves.

Despite her preacher's tone, Amelia Lewis was motivated by philanthropy and used the technologies of stove building and the new mass printing methods to drive her attempt to improve the lot of the poor. Despite the prodigious output of one decade, information about her before and after the 1870s is unfortunately sparse.

CHAPTER SEVEN

Transport, Travel and Technology

The transport revolutions brought by the railways in the 1830s and the bicycle in the 1880s introduced unprecedented freedom and independence to women as, at last, the means of travelling alone, wherever and whenever they wanted, were accessible. The repercussions were enormous as self-confidence permeated all aspects of their lives. Over these years escape from the constraints imposed on them for generations became a possibility and the world was there to be explored. As in so much else which affected them personally, women quickly devised all types of gadgets, accessories and improvements to meet these new requirements and aspirations.

Until George Stephenson ran his engine *Locomotion* on the line from Stockton to Darlington in 1825, travel over land had been entirely dependent upon the horse; meanwhile, freight was often carried by barges navigating the inland waterways. Further afield, intrepid sailors had discovered distant lands, fought wars and brought back the spoils of their travels on ships powered at the behest of the winds and guided by the stars. The first half of the nineteenth century would see dramatic changes to the landscape brought about by the railways and the need for bridges to cross their tracks. Various methods to suspend bridges over rivers and valleys had been used around the world for centuries. In Britain the era of the great bridge engineers and builders, such as Thomas Telford and Isambard Kingdom Brunel, had not yet begun, neither had the mass building of the railways. But new structural engineering practices, some influenced by the rigging on ships and others by the possibility of using large quantities of iron, were bringing striking new variations to bridge building. In 1779 Abraham Darby and Thomas Pritchard had built the first cast-iron bridge across the River Severn, upstream from its estuary at Bristol, in Coalbrookdale. In 1811 Sarah Guppy, the multi-talented wife of Samuel, a Bristol merchant, patented her method for 'Bridges and Railroads';[1] a year later she would register her combined egg cooker, toaster and coffee urn, and in the 1830s the magnificent bed with steps up to it was her final innovation. Her proposal of improvements for a suspension bridge is notable in its timing because Sarah Guppy receives no mention in the forthcoming era of the bridge builder. Thomas Telford began work on the Menai suspension bridge linking Wales and Anglesey in 1818,[2] seven years after Sarah Guppy's invention. Years later, when Brunel designed the Clifton suspension bridge across the Avon Gorge in Bristol, another Bristol merchant, Thomas Guppy, would be one of his investors. Yet, in her 1811 patent, Sarah Guppy describes a method to erect sturdy piles from which to suspend a bridge:

> On each side of the river or place over which a bridge or road is to be constructed, pursuant to my said invention, I do fix or drive a row of piles, with suitable framing to connect them together, and behind these I do fix, or drive, and connect, other piles or rows of piles and suitable framing, or otherwise, upon the banks of the said river or place, I do dispose or build certain masses of connected masonry or other ponderous structure, with piles or without, in order and to the end that the said piles or masonry, or other structures, shall be capable of sustaining and permanently resisting the action considerable force applied or exerted in directions tending to bring the same together; and I pass across the said river or place, from the upper or other convenient part of the said piles or masonry or structure, several strong metallick chains, parallel to and at suitable distances from each other, which said chains may be drawn tight by secure mechanical means; or otherwise the said chains may be suffered to hang in similar lines, slightly curved from the side or bank to the other, and in either case I do dispose upon the said chains longitudinally and crosswise such for pieces of timber, or iron, or other suitable material, as shall constitute a platform, which, by the connection or disposition of the materials thereof, shall afford a proper support for a road or pavement of the usual structure, or for railroads, which last, namely, the railroads, upon such occasions as may require the use and application of my said Invention, I do connect, unite, and frame together with each other and with the claims herein-before mentioned and described.

Despite her appetite for invention and her ability to focus on the smallest and largest of concepts, Sarah Guppy's name does not appear in the histories of engineering and bridge building.

The rapid expansion of a rail network across Britain in the 1830s was received with interest, excitement and trepidation. Some of the first passengers thought they were flying, others were enamoured with the novel experience and more were terrified by the noise and smoke. Quickly, track was laid, stations designed, bridges and viaducts built and tunnels excavated as towns and cities were connected to each other. In 1838, Euston station opened in London and the line was laid to Birmingham; another ran between Dundee and Arbroath; and Brunel designed Paddington station for his Great Western Railway to

Maidenhead, taking it on to Bristol in 1841. Not only did the railways bring mass transport facilities and independent travel to Britain, but also many lucrative commercial openings ready to be exploited – the Royal Mail, cables and telegraphy, package holidays, news agencies – all of which were to have profound effects on women's lives as consumers and designers. Entrepreneurs like the travel agent Thomas Cook, the international journalist Baron Paul Reuter and a newspaper and bookseller, Mr W.H. Smith, established successful businesses.

For women, perhaps the most important aspect of the railways was the newfound independence it gave them. Writers like Charlotte Brontë used their image as a metaphor for women's opportunity for escape and excitement, even to go abroad. She wrote for Charles Dickens's weekly magazine, *Household Words* which was sold along with journals like the *Englishwoman's Domestic Magazine*, on one of Mr Smith's station bookstalls. With the recent advances in printing technology, and the correct recognition that people would wish to read on their journeys, his bookstalls became highly lucrative ventures. Soon after she had seen a train for the first time, Queen Victoria ordered a royal train to take her and the family on the GWR to Windsor. In contrast, for thousands of young women, the railway was a means by which they left their rural homes and families to travel to their new lives in domestic service.

Thomas Cook, who had organised train journeys to temperance meetings in the Midlands, introduced, in 1851, the inclusive excursion ticket, purchased before departure, offering a return trip, overnight stay and entrance fee to the Great Exhibition in Hyde Park. He quickly built on his success and ventured on to the continent with holidays based on routes through France, Germany and over the Alps into Switzerland and on to Italy. Suddenly the Grand Tour, which for decades had been a rite of passage for upper-class young men, became affordable to middle-class women and men. Manuals appeared specifically for women with titles like, *A Few Words of Advice on Travelling and its Requirements: Addressed to Ladies by a Lady*. This book, published in 1875, contained information on the new code of behaviour, routes, seasonal clothing, luggage, personal safety, souvenirs and postage as well as French and German traveller's dictionaries.[3] Some women found work on the railways, providing beverages at the stations, running the ladies' waiting rooms and sometimes taking charge as station mistresses. Mrs Argyle ran the station and signals at Merrylees on the Midland Railway line for forty years and Mrs Huxley was mistress of Braceborough Spa station in Lincolnshire.[4]

Although the ladies' waiting rooms on the station platforms allowed women a place to rest away from men, they did travel together on the trains and this seemed to lead to all kinds of wanted

EXPRESS TRAINS

AT CONVENIENT INTERVALS BETWEEN

LONDON AND LIVERPOOL

(ST. PANCRAS). (CENTRAL).

** The MIDLAND is the ONLY LINE between LONDON and LIVERPOOL passing through the Magnificent and Picturesque Scenery of the Peak of Derbyshire and the Vale of Matlock.*

SPECIAL EXPRESS TRAINS are run between LONDON (St. Pancras) and LIVERPOOL (Central) for a reasonable number of Passengers when required in connection with American Steamers.

DRAWING ROOM SALOON CARS by the Day Express Trains between London and Liverpool.—Holders of First Class Tickets, viâ the MIDLAND RAILWAY, can use these Cars WITHOUT EXTRA CHARGE.

PRIVATE DRAWING ROOM SALOONS with LAVATORY and other conveniences are provided for the exclusive use of parties of SEVEN or more WITHOUT EXTRA CHARGE.

EXPRESS TRAIN SERVICES BETWEEN

LONDON (St. Pancras) and **SCOTLAND**; also between **LIVERPOOL** (Exchange) and **SCOTLAND**; the Direct Route to and from

Glasgow & Greenock (for the **Western Highlands & Islands**), through the **Land of Burns**; **Edinburgh**, through **Melrose** and the **Waverley** District; **Perth**, **Aberdeen**, **Inverness**, &c., over the **Forth Bridge**.

☞ *The opening of the Forth Bridge has materially shortened the journey between the Midland System and North Scotland.*

THROUGH TRAIN SERVICES also between LIVERPOOL and Principal Towns in the MIDLAND COUNTIES and WEST OF ENGLAND.

THROUGH CARRIAGES of the newest and most improved pattern, provided with FIRST AND THIRD CLASS LAVATORY ACCOMMODATION, by the Midland Express Trains.

BAGGAGE IS CHECKED THROUGH from London (St. Pancras) to Liverpool, and delivered to Tender, or Steamer if it leaves direct from the DOCK; also from New York or the Landing Stage at Liverpool to any Hotel, private residence, or Railway Station in London.

LUNCHEON BASKETS may be obtained at many of the principal Midland Stations, and delivered at the train *en route* on application.

The "Adelphi Hotel," Liverpool (ADELPHI HOTEL COMPANY), reorganised, refurnished, and redecorated, is now **one of the best of European Hotels.**

The Midland Grand Hotel, attached to the LONDON (St. Pancras) STATION, is one of the **Largest and Best Appointed in Europe.**

Every information required by travellers may be obtained from Mr. J. ELLIOTT, London (St. Pancras); Mr. JOHN B. CURTIS, 21, Castle Street, Liverpool; Mr. M. H. HURLEY, the Company's American Agent, 261 and 262, Broadway, New York; or Mr. W. L. MUGLISTON, Superintendent of the Line, Derby; also from Messrs. THOS. COOK AND SON, Official Agents of the Midland Railway Company, and Sole Passenger Agents to the Royal British Commission for the Chicago Exhibition.

☞ **A Show Case, containing Photographs of Places of Historic and Scenic Interest on or accessible from the Midland Railway, is on view in the Transportation Building, to which the attention of Visitors is invited.**

Derby, 1893. GEO. H. TURNER, *General Manager.*

Advertisement for the Midland Railway from the *Official Catalogue of the British Section*, Royal Commission for the Chicago Exhibition, 1893. (*Author's Collection*)

and unwanted relationships. Not only was a woman in charge of her independence but also of her own safety. One can only speculate whether in 1904 Louisa Llewellin, a lady from Holloway in London, had had the opportunity to use her 'Gloves for Self Defence and other purposes'.[5] Miss Llewellin had scant regard for the injuries the user might inflict upon her attacker and only a determination that she should overpower him and make him identifiable. These gloves could be worn for the whole or part of the journey; they had sharp steel talons or nails at the ends of the fingers and were for

The Indiscretion, by Constant Aime Marie Cap. (*Berko Fine Paintings, Knokke-Zoute, Belgium/Bridgeman Art Library*)

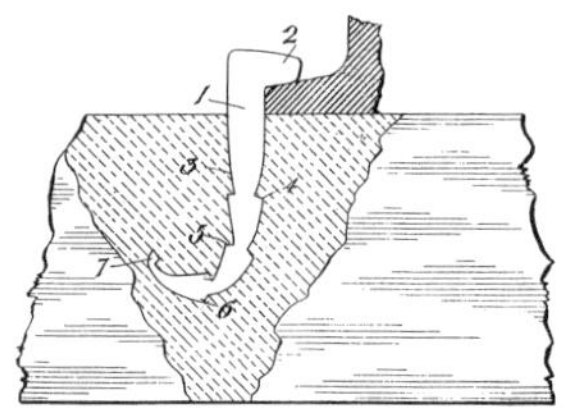

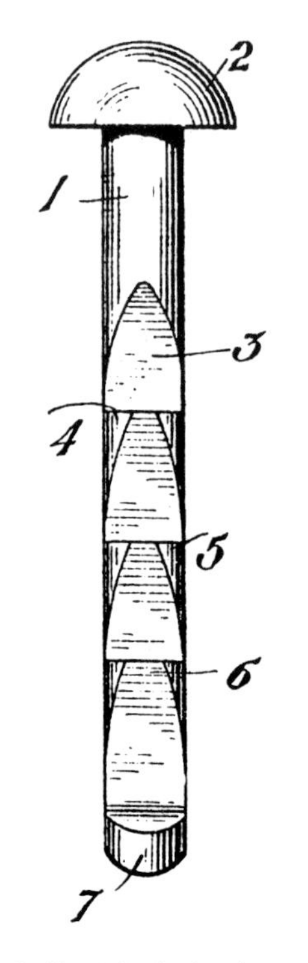

Eva Balfour's barbed spikes to secure sleepers on to the railway track. Patent.

> self-defence and other purposes and more especially for the use of ladies who travel alone and are therefore liable to be assailed by thieves and others. The object is to provide means whereby a person's face can be effectually disfigured and the display of the article which forms the subject of my invention would speedily warn an assailant of what he might expect should he not desist from pursuing his evil designs, and the fact that he would in the case of persistence be sure to receive marks which would make him a noticeable figure would act as a deterrent . . . and be so severely scratched as to effectually prevent the majority of people from continuing their molestations.

Eva Balfour from Bromley in Kent was more interested in security of the railway track. She introduced a barbed spike with a large head to secure railway sleepers permanently on to the track.[6] The ratchets on its sides and the ability to curve inwards once it was banged into place meant that it could not be removed. The sleeper and track were therefore securely based, which was essential in order to avoid some of the dreadful derailments and accidents on the railways.

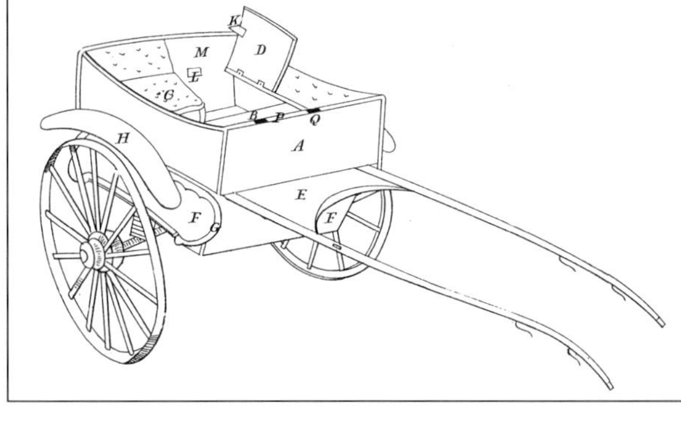

The rein-bearer and driver's seat could be adjusted to the height of the horse in Frances Young's carriage. Patent.

Horses continued to provide a vital form of transport throughout the nineteenth century. Frances Young from Norwich made modifications to the driving of an open carriage[7] by adapting the rein-bearer to suit the height of the horse and making the driver's seat adjustable to enable better control of the carriage. Even as late as 1907, by which time there were many patents for special clothing for bicycle riding, Mabel Van Vechten, a nurse from New York City, contrived an improvement to side-saddle riding.[8] She made a foot piece, based on a reversed stirrup, in which the right foot was now firmly in place making it easier for the rider to rise when trotting.

Few could have predicted the ensuing communications revolution when, in 1837, a cable was placed alongside the railway line from Euston to Camden Town in London. The purpose of this exercise was to prove that sound could be transmitted along a wire beside the track, and so telegraphic communication around the world using cables began. A national postal service became possible with the railway network and the Penny Post was introduced in 1840. The Crimean War (1854–5) was the first war to be recorded by journalists and photographers on the field. Reports reached Britain within a few days, having been cabled across Europe to London. The new techniques of photography enabled people like Roger Fenton to supply the press with up-to-date pictures from the theatre of war. At home, wives and mothers were able to read about and see visual images of the horrors, mistakes and successes of the military campaign against Russia on the shores of the Black Sea. A doctored account from government sources produced weeks later, perpetuating

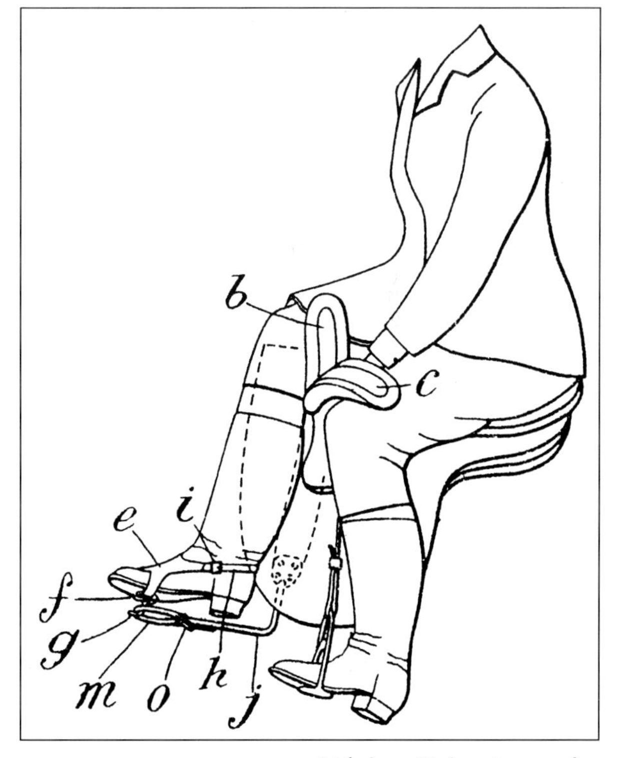

Mabel van Vechten improved foot stirrups to enable women to ride side-saddle more easily. Patent.

EXHIBITED IN BRITISH SECTION,
MANUFACTURES BUILDING.

THE "TWIN ZENITH" PATENT

SAFETY COMBINATION RIDING HABIT.

OPINIONS OF THE PRESS.

"All the more serious accidents to ladies have been caused from their inability to clear the skirt from the pommel in a fall. No such difficulty could occur with Mr. Shingleton's patent; hence it is to be hoped it may be universally adopted."—*Court Circular.*

"It is with genuine pleasure that we can at last record the invention of what may honestly be called a safety habit. Hitherto all attempts at this much-to-be-desired garment have been either unsightly when out of the saddle, or have not fulfilled their purpose of securing safety to the wearer, if not life. The 'Twin Zenith' carries out both intentions. While in the saddle it has the appearance of an ordinary habit; on dismounting, the adjustment of a single button removes all possibility of unsightliness with its attendant discomfort. Mr. Shingleton's invention combines breeches and skirt in one."—*Land and Water.*

"The safety skirt of the age."—*Black and White.*

"Mr. Shingleton's new invention is an admirable contrivance, and will ensure those women who adopt it from the most alarming of all accidents that can befall a horsewoman."—*Queen.*

This Patent has been tested by a number of well-known Ladies in the hunting field, four of whom have been thrown, and pronounce as to its safety. Testimonials can be seen.

W. SHINGLETON, Tailor and Habit Maker,
60, NEW BOND STREET, LONDON, W.
ARTHUR DAY, Representative at the Exhibition.
THE AMERICAN PATENT FOR DISPOSAL.

Even in the 1890s women still rode side-saddle, as this advertisement for the 'Safety Combination Riding Habit' shows, from the *Official Catalogue of the British Section*, Royal Commission for the Chicago Exhibition, 1893. (*Author's Collection*)

heroism and mythical victories, was no longer acceptable. Women were able to see exactly what their men were involved in; maybe at this time a soldier lost some of his glamour. A successful empire was dependent upon communication, and the combination of steam-powered ships, railways and now telegraphy was a major influence in the expansion of British imperialism.

Telegraphy and two other inventions – the typewriter and the bicycle – introduced whole new areas of activity highly appropriate

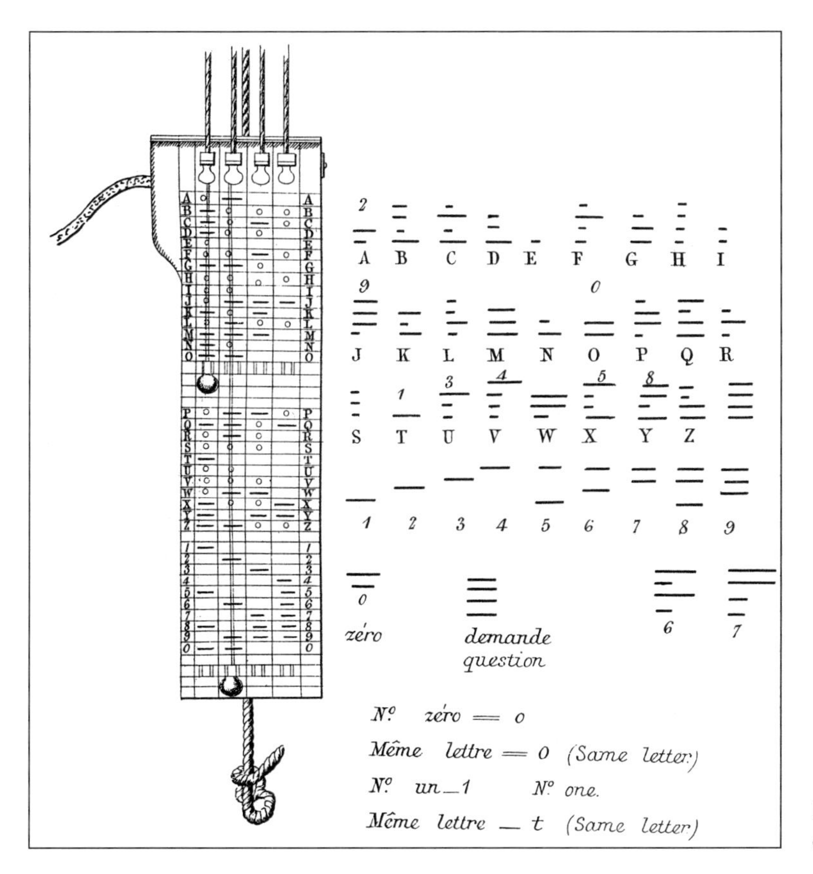

Part of Sarah Buckwell's coding system. Patent.

for women and further enhancing their independence and self-esteem. After the Married Woman's Property Act of 1884 even married women could legally retain their own earnings. When it was found that women were more patient and polite than men, and that their voices carried better, thousands of them were employed in the new telegraph exchanges.[9] So a new profession emerged, and in the 1901 census, many gave their occupation as 'Telegraphist, General Post Office'.[10]

Women were encouraged to join their husbands on colonial service in India, and whole families were enticed to emigrate to other colonies by the government, eager to populate Canada, Australia and New Zealand, with promises of larger houses, fresh air and opportunities for work. Safety at sea became a preoccupation for many, especially adequate signalling and methods to save lives. Samuel Morse had introduced his code in 1838 using the new telegraphy and his method of communicating with dots and dashes became widely used. Clearly influenced by Morse's code, Sarah Buckwell, one of a growing number of British women living in Europe, patented her invention for 'Telegraphy' from Luino in Italy in 1865.[11] In her system, intelligence was communicated by a series of movable plain or coloured diaphragms, signs and symbols on blinds or shutters, each representing letters. The method could also be used to transmit messages: 'the type may be set up in a line. Or arranged spirally round a cylinder, and caused to traverse the current breaker, so as to work it, and transmit currents through the wire of such duration and at such intervals as correspond with the message transmitted'. The system was adaptable: sounds made with a hammer or gong could also be used, the combination of different lengths of sounds representing the different letters. In foggy weather it could activate signals that would launch coloured rockets or flares, again each representing a different part of the code.

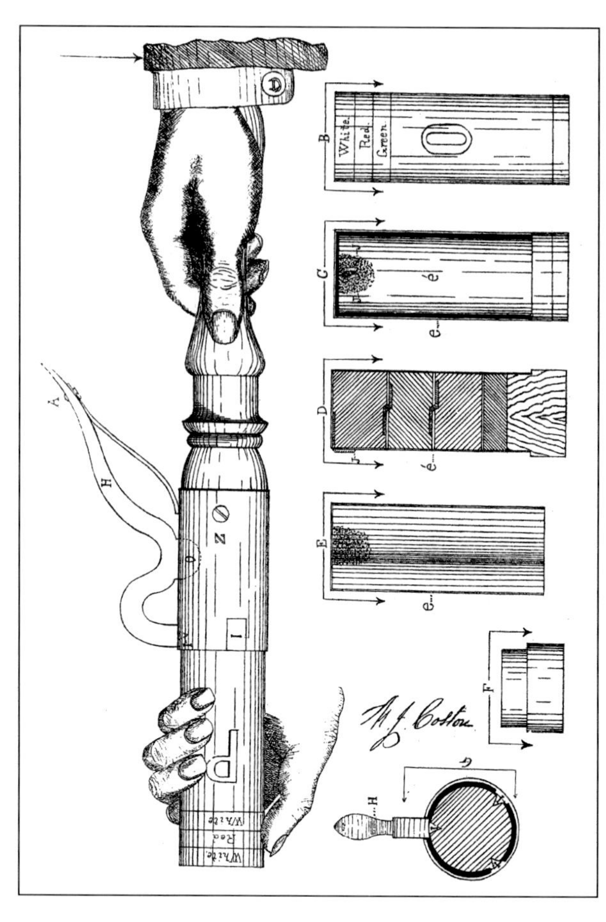

Martha Coston's Improvements in Pyrotechnic Night Signals, with her signature as their inventor. Patent.

A comparable idea was put forward in the United States by Martha Coston, whose husband had died leaving her, aged twenty-one, with four young children to support. While going through his papers, she discovered a 'Method of Signalizing Any Numeral or Combination of Numerals by the Display of Different Pyrotechnic Fires'. This system of coloured signal flares was intended for use at sea and, importantly, at night. Martha Coston patented her husband's invention in Britain and the USA in 1859,[12] acknowledging Franklin Coston as the originator of the idea. Effective use of gunpowder at sea, especially in the dark, was very difficult to achieve. Coston's was in three colours and packed into different sized boxes that had to be kept waterproof for successful ignition. Different colours and lengths of flare corresponded to different numbers, for example, one was represented by a single, short white flare and eight was a blue flare running into a red one. A lengthy white flare meant that a person on the ship needed to communicate with another ship using the Coston signals; in response a long red flare indicated that the message had been received and understood. The relevant flare number could be cross-referenced in an already existing codebook. However, there were problems with this method, which Martha eventually resolved over the following twelve years. The major improvements in her signals were that they could be hand-held and were contained in an ingenious

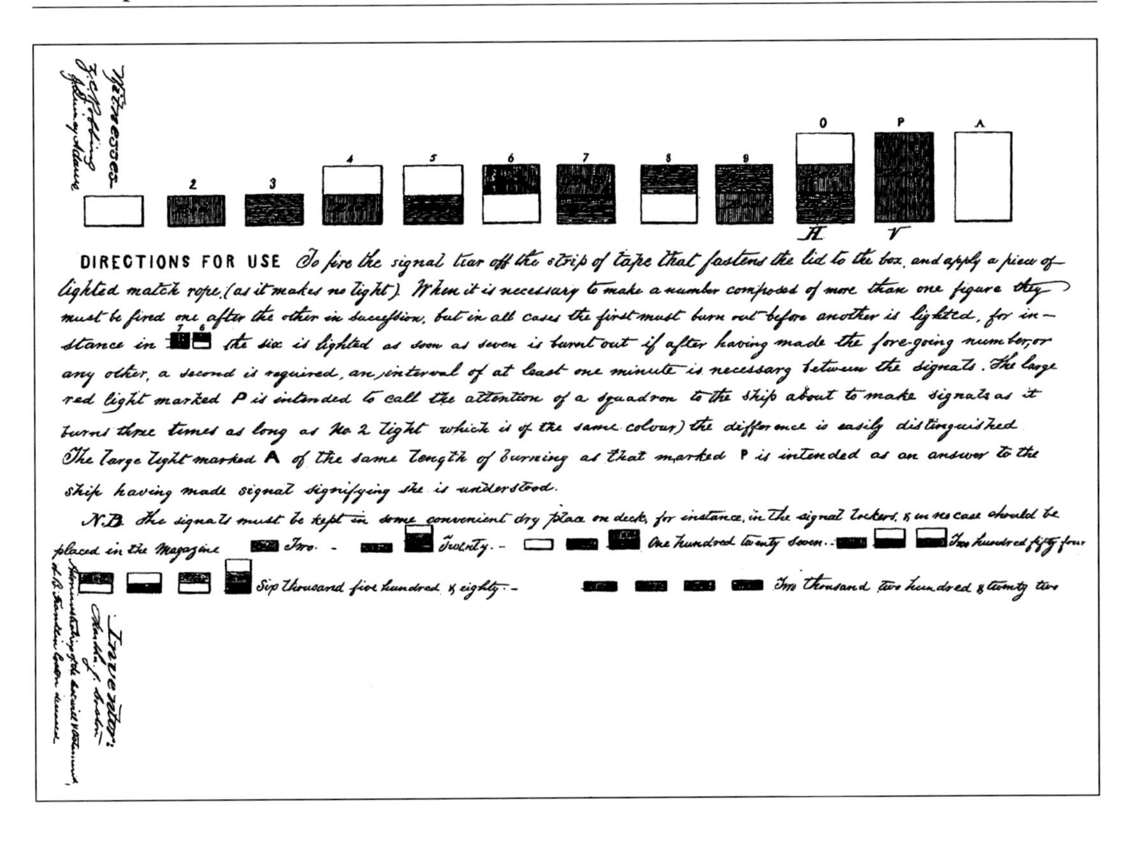

Instructions for using the signals. Patent.

self-igniting cylindrical case. When the outer part of the case was twisted over the inner, it caused enough friction for a 'quick-match' to ignite the gunpowder. Her instructions state clearly how to set off a flare, speed and keeping dry being essential when at night, in the dark, on a sinking ship:

> Grasp the holder in the right hand and insert in its top that end of the signal that has no coloured stripes . . . Lift a little the spring II, that the steel point I of the same may penetrate well into the wooden bottom F. Then grasp around the upper part of the signal with the left hand . . . firmly holding the holder in the right hand, and give the signal holder a twist in opposite directions . . . The ignition announces itself with a fizzing sound. The twisting of the signal loosens the outer case E, which surrounds the signal proper and produces a friction. Raise off this outer case E and the ignition is complete, when the signal continues to burn, showing colors as numbered or signified by stripes of colored paper.

Eventually Martha Coston sold her signals to the US navy and to shipping companies with ocean-going liners. But the navy tried to claim the idea was theirs: an action that Martha Coston fought and

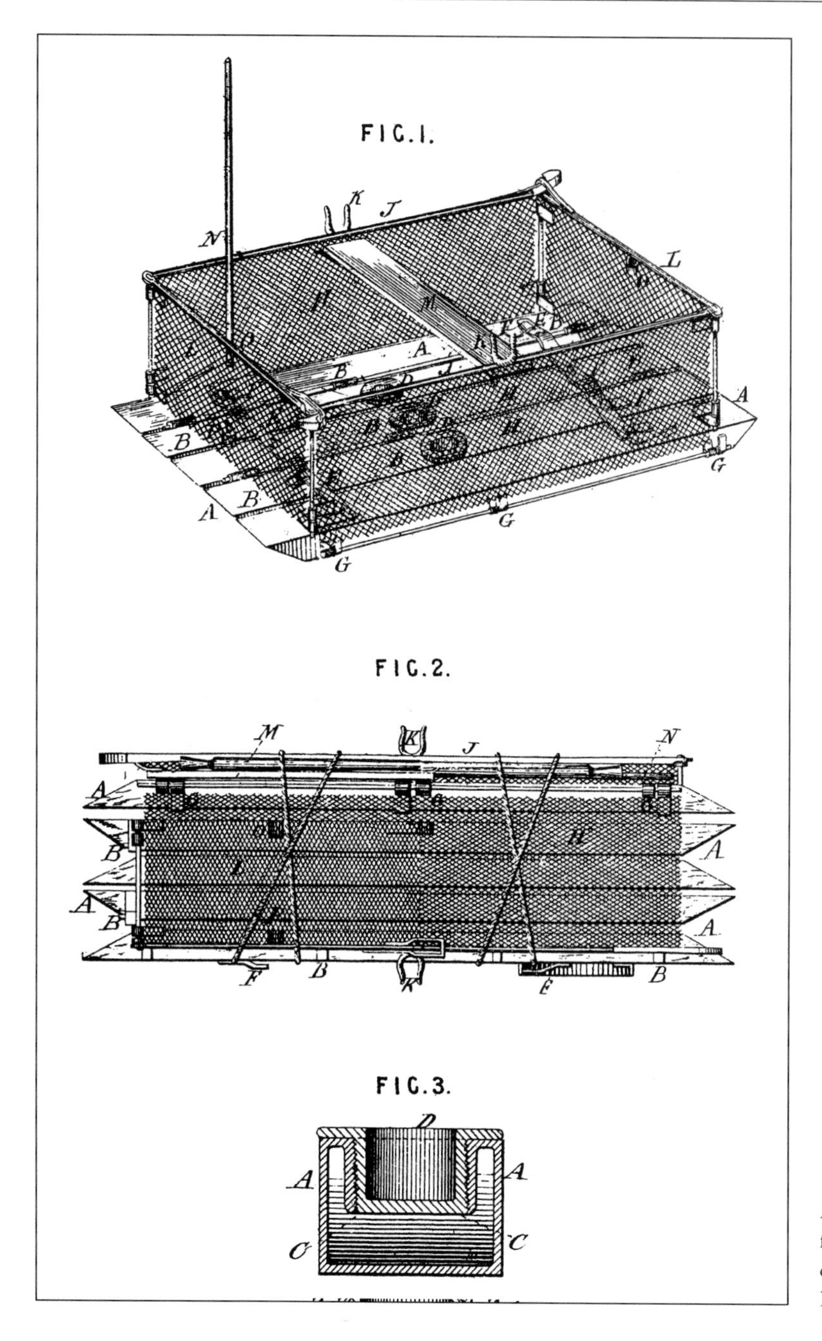

A folding, metal, fireproof raft for use in case of shipwreck designed by Maria Beasly. Patent.

won in 1871, when she patented her 'Pyrotechnic Night Signals' in Washington.[13] With her invention Martha Coston did eventually make money and, more importantly, thousands of lives were saved.[14]

A shipwreck at any time must be one of the most frightening and traumatic of experiences. While most ships carried lifeboats, evacuation of passengers on to a sometimes ill-equipped boat or raft did not necessarily mean survival. There were the looming threats of drowning

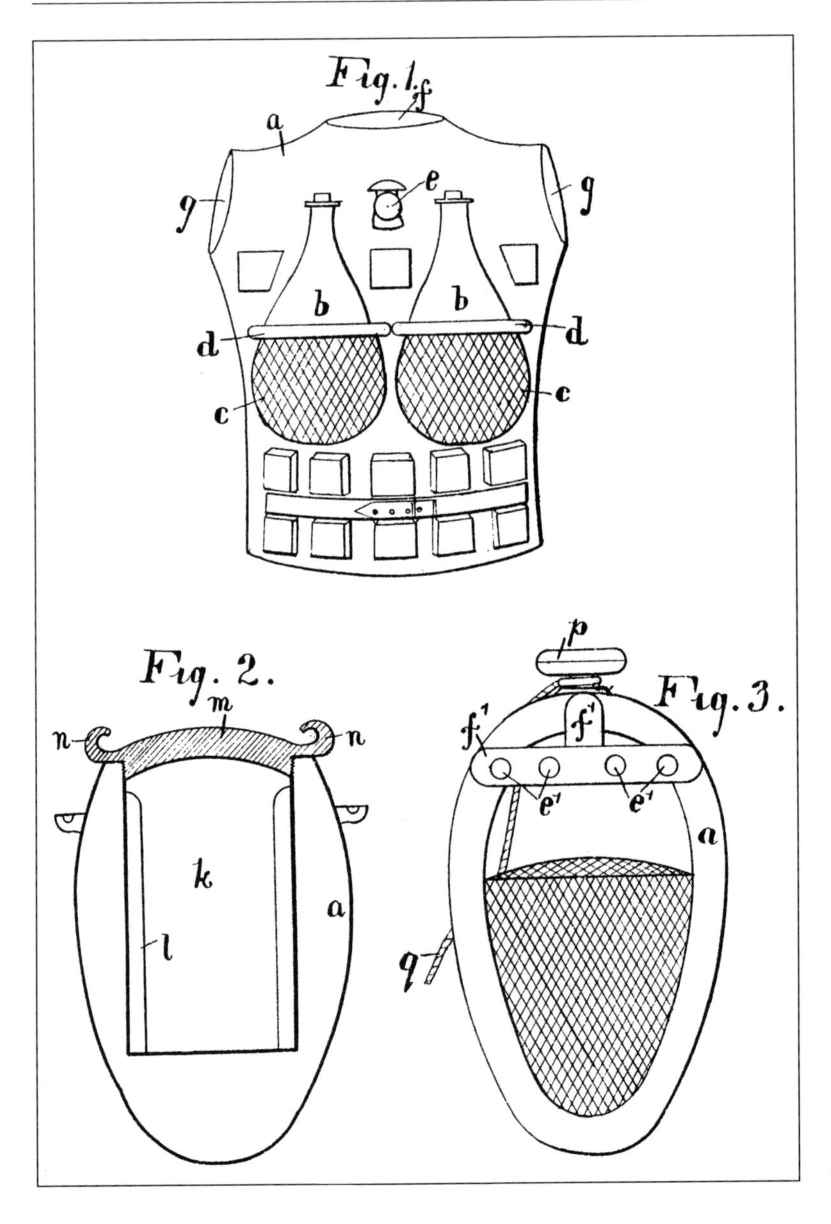

Kate Jenkins invented this life jacket that could also be used as a floating buoy. Patent.

and death by fire, as well as the cold, hunger and exposure to stormy weather stranded in the middle of the ocean. Maria Beasly from Philadelphia was aware of all these hazards when she introduced a folding, metal, fireproof raft in 1880,[15] made of a series of hollow metal rectangular floats, which were hinged together to form a platform when unfolded. The side could be enclosed with bars or railings and there were supports for oars and storage spaces for emergency rations which were to be 'kept continuously ready for any emergency of shipwreck'.

Two years before the sinking of the *Titanic* in 1911, Kate Jenkins from Sydney, Australia, equally conscious of the horrors of shipwreck,

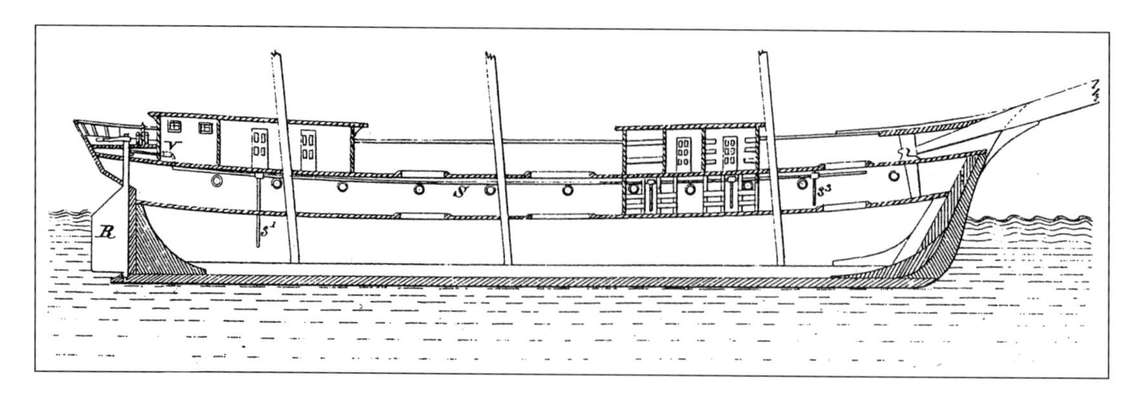

invented her life jacket which could also double as a floating buoy.[16] The jacket was made from cork slabs, fastened with brass rivets and lined with a material to retain body heat. Netting pockets on the back and front held containers of water, wrapped in rubber tubing, inflated with air for buoyancy. A small phosphorous light was placed in the middle of the chest. Other watertight pockets were for food, necessary supplies, important papers, jewellery and money. When the same principles were converted into a buoy, the top was covered in phosphorescent lights and a large pocket might contain the ship's papers. On the very top was a small metal pocket in which a last message could be left and a rope on the side acted as a lifeline to tow it ashore.

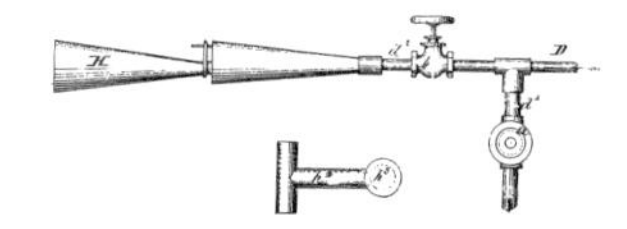

Top: The electrician Amelia Pitts Armstrong and her colleagues designed a ventilation and foghorn system for ships. Patent. *Above:* Part of the foghorn that was inserted into the exhaust end of the ventilation system. Patent.

Shipwreck apart, conditions on board ships were appalling, especially for the majority: those living and working below deck. Amelia Pitts Armstrong, a New York electrician, and her colleagues invented a mechanism to ventilate ships, deodorise and disinfect 'the disease – germs and any noxious odours' emanating from them. It could also act as a foghorn.[17] The description of the necessity for such devices reveals horrific details of life on board, and the conditions endured by those travelling below deck in the bowels of the ship. The stench from humans and animals, rotting cargoes and the generally unsanitary conditions of the thousands travelling in steerage must have been almost unbearable. Amelia Armstrong's invention used a tubular rudder or tubes attached to it; the action of the water on it would pump the air entering through a valve into the pipes and pass it around the ship through a series of automatic suction and discharge valves. These could be regulated to discharge the appropriate amount of fresh air through vents. A foghorn could be attached at the end of the piping so that the exhausted air could pass through it and make a sound if necessary. Deeper in the ship this ventilation became vital to

> effectively remove from the lowest depths and the most remote recesses of the hold and steerage the heavy foul air arising from bilge vapor and the decay of animal and vegetable matter, and the heated air and explosive gases from cargoes . . . floating morbific germs of infectious or contagious diseases, such as Asiatic cholera, yellow fever, ship-fever, small-pox.

Rather than exhale the germ-laden air from the ship into the atmosphere, where the wind could blow it back again or on to another ship, it could be pumped into a separate chamber of chemicals where it would be heated and disinfected, then discharged.

An advertisement for Cadbury's Cocoa showing the deck aboard an ocean-going ship. *(Illustrated London News)*

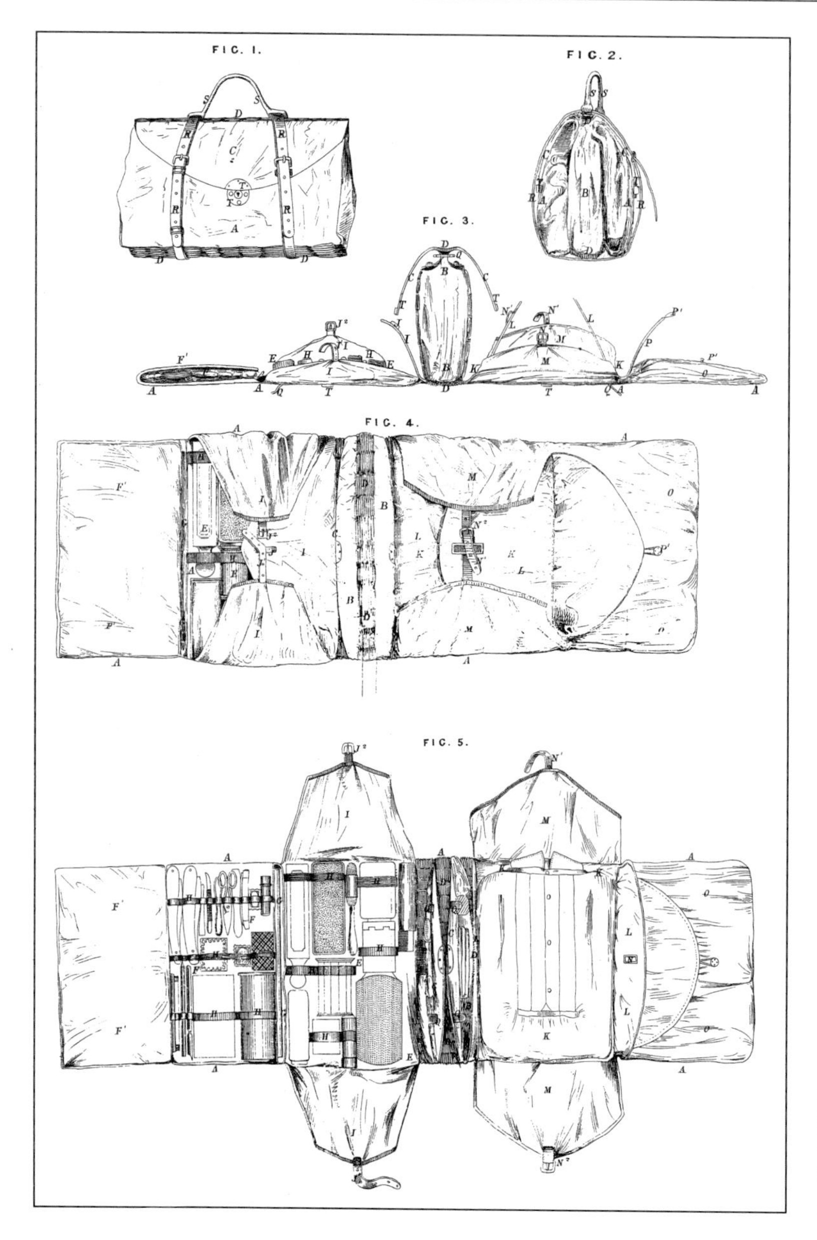

All necessities for a long journey could be safely packed and stored in Charlotte Mansel's 'United Service Travelling Case'. Patent.

While survival was the highest priority on these long voyages, there were many inventions, particularly for those travelling on the upper decks, to help cope with life on board on a day-to-day basis. In 1859 Charlotte Mansel,[18] from the naval town of Plymouth, designed a wonderful bag made of a long length of oilskin or leather. Copious pockets and numerous elasticised straps were fastened to it in which to put all the odds and ends required for travel. She called it 'The United Service Travelling Case' and recommended it 'for carrying linen, toilet apparatus, stationery, food, spirit flasks, sandwich and provision boxes, looking glasses, paper and writing pads . . . and any other articles usually required by travellers'. The whole thing was then neatly wrapped up, strapped together and carried with a handle. Secrecy on a journey was uppermost in Mary Boyle's mind when she designed her portable writing case and despatch box in 1863.[19] She lived in Kensington, which was then the wealthiest of London's boroughs, and no doubt its resident diplomats and city speculators and traders, requiring confidentiality in their dealings, were among her friends. Her invention had secret drawers and panels, a paper holder and space for books. When closed, a space between the desk and the fittings contained pens and ink. Before the invention of Laszlo Biro's ballpoint pen, writing on a journey was potentially very messy: inkbottles could be knocked over and clothes ruined. Elizabeth Stiles from Hartford, Connecticut in the USA designed self-regulating inkwells for use on a rolling ship or in a swaying carriage.[20]

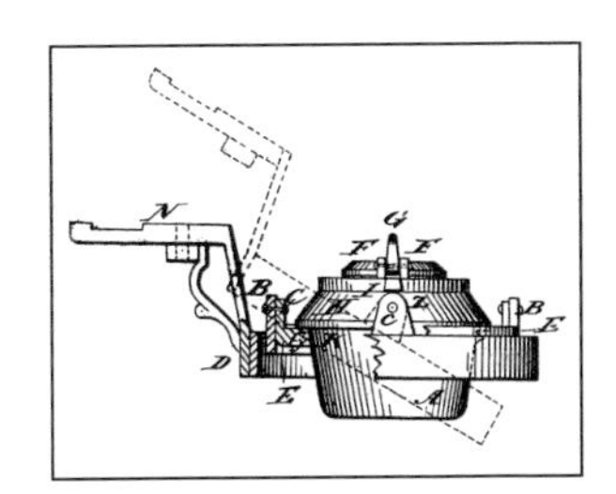

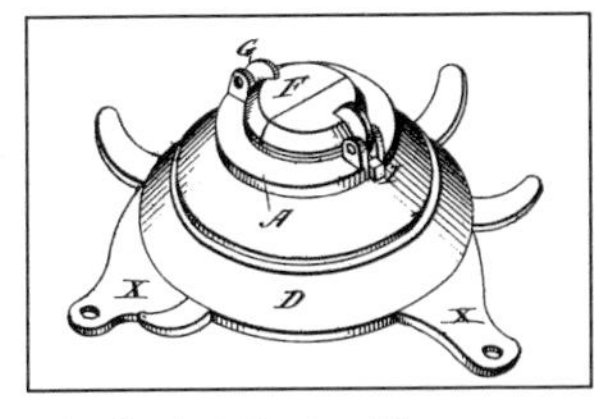

Elizabeth Stiles's self-regulating inkwells for use aboard ship. Patent.

Aware of the need to keep her children occupied on the voyage, Elizabeth Kelley from West London devised a musical instrument[21] which could be used for entertainment and to 'replace the pianoforte for children's practice on board of ship'. It consisted of 'a case with a keyboard divided crosswise into two or three sections joined by hinges, surrounded by borders'. Beneath the keys was a cavity to house 'the mechanism of the interior, the tones of which are produced by metal pieces, chimes, bells, wire, catgut, vellum, glass and china'. While the sound was probably not that of a concert instrument, it no doubt did entertain its players.

The travelling trunk was subjected to all types of modifications and adaptations for use both on the journey and in the new home. Mary Cecil Gladstone, from Great Barton Vicarage in Bury St Edmunds, exhibited hers at the Chicago Fair in 1893.[22] This was a leather or wicker trunk with two drawers, which were pulled out of the side. Wickerwork or leather were also used by Grace Maitland from Hampshire in 1913,[23] but this time to convert her trunk into a comfortable lounge chair which, no

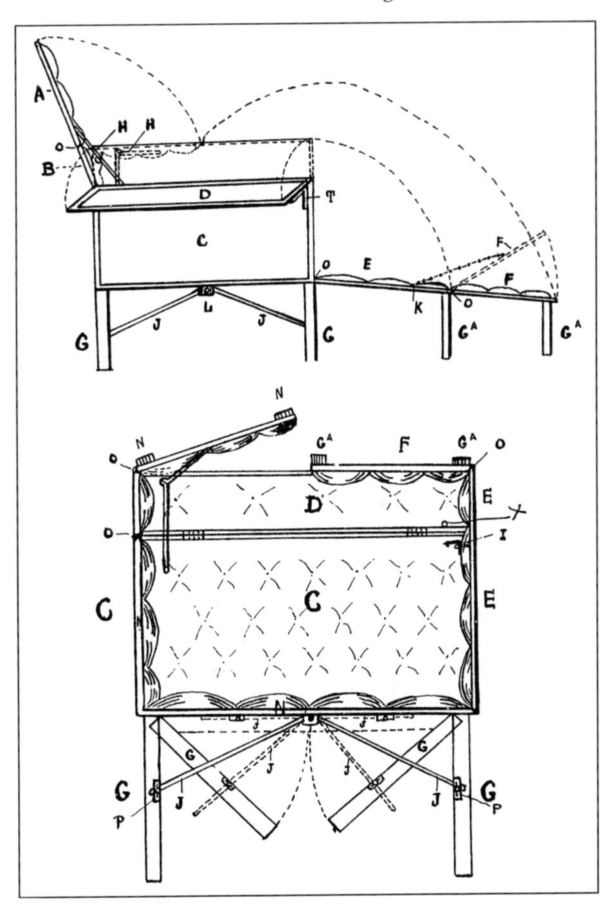

This dual-purpose travelling trunk, designed by Grace Maitland, converted into a reclining chair. Patent.

doubt, could have graced a first-class deck and then become part of the furniture in a new home. It 'was constructed so as to be useful on a liner for deck chair, or drawing-room rest-chair. Travellers to the colonies often discover that trunks though useful and necessary for carrying clothes or any other article, are very much in the way when they are settled in a house or flat.' Lined with canvas, the top and front sides of Grace Maitland's trunk opened out to make a back and footrest and the whole thing stood on short legs. The work of Maria Monzani and her husband echoed this need for flexible and portable furniture in the 1850s. There was also a travel cot designed by 28-year-old Anna Raw from the port of Liverpool in 1907 and no doubt used on many of the ocean liners full of emigrants departing to the United States.[24] This cot was made of fabric hung over a collapsible wooden frame, with bolted corners making it quick and easy to assemble.

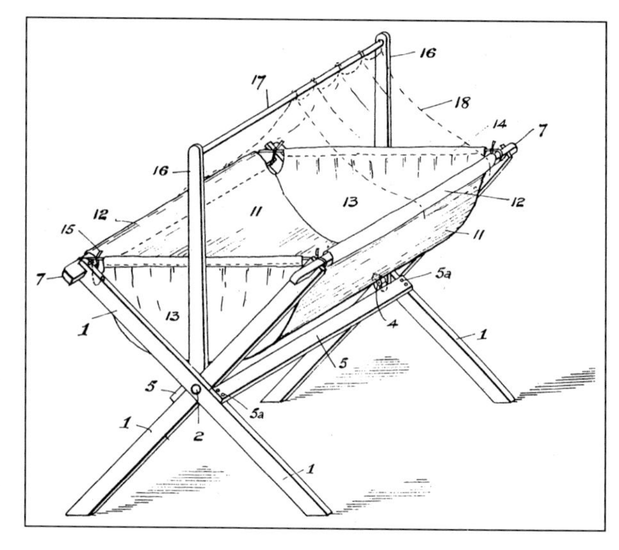

A simple folding wood and canvas cot, by Anna Raw. Patent.

Even with the journey over, life in a tropical climate, before immunisation, could be problematic, as implied by Sarah Beauchamp Bush in 1870[25] in her 'Improved Means of and Apparatus for Preventing the Entrance of Flies and other Insects and Reptiles into Rooms or Apartments of Buildings, Ships and Carriages, Sleeping Berths or other Places . . . by means of metallic blinds, curtains, screens or covers of plain ornamental character'. Her method to make a mosquito net provided ample scope for the needlewoman to incorporate her own designs and motifs. In it she would take

> very fine, gold, silver or other metallic wire and weave it into a suitable shape or form to act as a curtain, blind or screen . . . which may be stretched in a frame rolled upon rollers so as to be removed when not required . . . or used as a loose covering to be thrown over a bed . . . to protect the occupants from attack and annoyance of flies, mosquitoes and other insects, as well as reptiles and small animals. The wire may have silk or other threads woven with it to form designs or patterns . . . or very thin metallic plates or sheets . . . being perforated, pierced or punched . . . to admit light and ventilation. The perforations or other openings may be arranged to form geometrical figures or designs. If the curtains or blinds are framed or rolled upon rollers they can be used in lieu of glass in the panels, doors and roofs of palanquins, howdahs, close carriages and buildings of all kinds as well as . . . ships' cabins.

Women at work in the new telephone exchange in Kensington, 1902. (*Illustrated London News*)

By the 1880s the new class of professional women, going out to work at the telegraph exchange or, from 1873, in one of the offices equipped with Mr Remington's new typewriters, were taking advantage of the latest novelty, the bicycle. Not only did the bicycle give them more independence but also introduced a healthier lifestyle and, as has been seen by the loosening of their corsetry, a general reappraisal of their constricting Victorian dress. But there was disapproval from those appalled at the proposition of the 'feeble female' being replaced by this even more, independent and physically strong woman. Familiar warnings in the press cautioned that by loosening their corsets and pedalling bicycles women were going to give birth to small and sickly babies, thus threatening the evolution of the human race. But the contrary was true, as Patricia Marks reveals. By loosening their corsets they had larger, healthier babies and the consequent improvements to their physical well-being seemed to enhance their minds and intellects.[26] Another early obstacle to cycling concerned modesty and especially the danger of exposing the ankle to view. Margaret Corrie, from Woking in Surrey, addressed this challenge with her ankle guard or screen.[27] This she proposed to attach to the cycle frame 'for the sake of screening the feet and ankles of lady cyclists, and preventing them becoming objects of observation to individuals behind'.

An advertisement for Gladiator Cycles, Boulevard Montmartre, Paris. French School. (Barbara Singer/ *Bridgeman Art Library*)

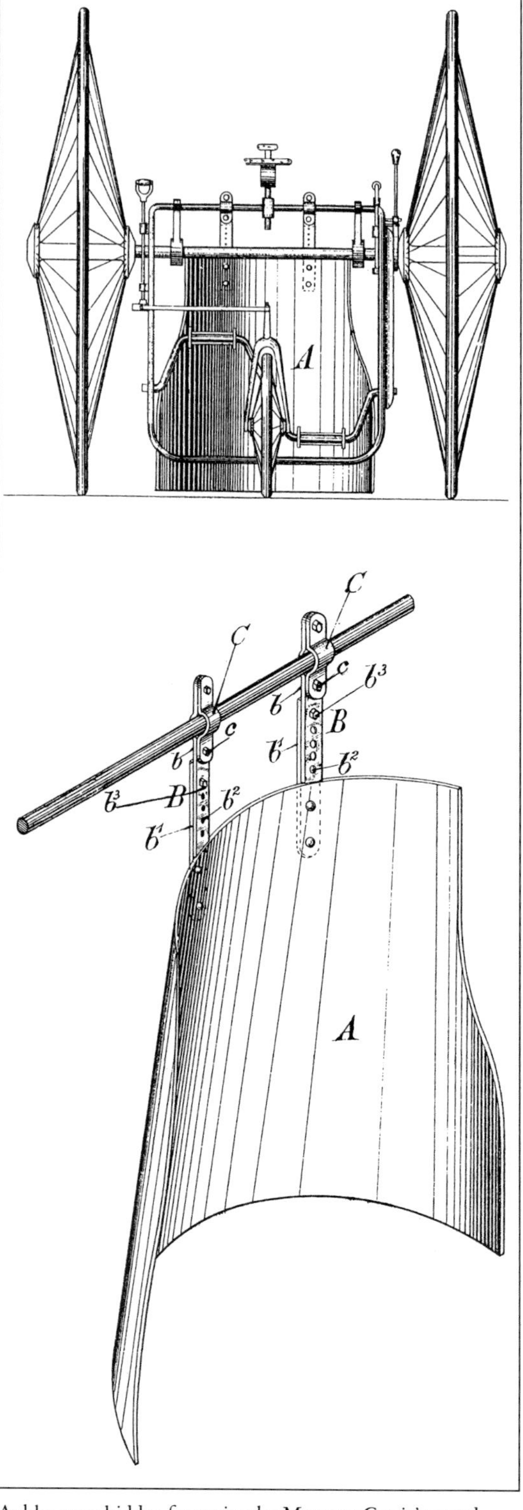

Ankles were hidden from view by Margaret Corrie's guard attached to the cycle frame. Patent.

Clarissa Jay's umbrella, which was fastened to the handlebars of a bicycle. Patent.

As has already been demonstrated, the boning of corsets now had to allow the hips freedom to move, the skirt to be divided and hats made to stay on the head in all weathers. A professional 'hair curler' from Yorkshire, Clara Moore,[28] designed an artificial fringe or curls, to be stitched or glued to a band and worn by a cyclist whose real hair might become uncurled or dishevelled.

Clarissa Jay of Adelaide, South Australia, probably as a consequence of the Australian extremes of weather, used an umbrella attached to her bicycle as a protection against sun, snow or rain.[29] More practically, Lady Mabel Lindsay from Abingdon designed a cycle rack on a cart to transport up to seven bicycles at a time.[30] Clearly an experienced cyclist and by profession an electrician, Lavinia Laxton designed an axle or spindle which would allow the rider to rest both feet on the pedals in the same position when not pedalling.[31]

One creative spinster, Clara Louisa Wells, has left quite a document of the product of her somewhat fertile imagination. From 1889, she spent at least ten years at various locations in the south of France and Italy, giving her address as care of a local British business or consul in each place. She was clearly influenced by the new technologies; their possibilities and her inventions seemed to combine with inspiration from the specific geography and geology of

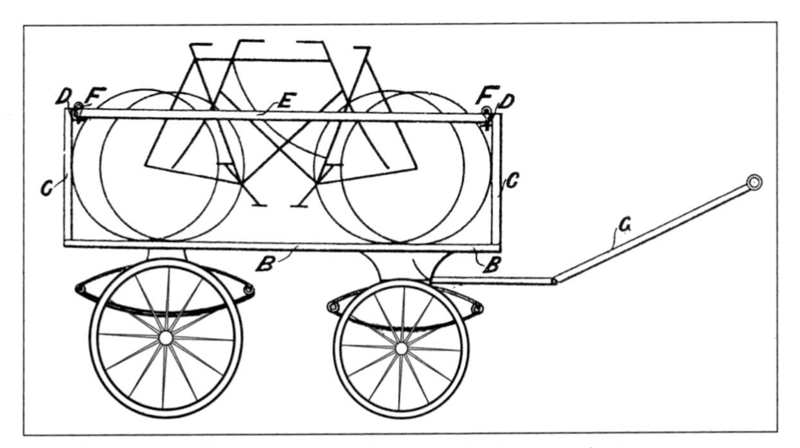

Seven bicycles could be transported in Mabel Lindsay's cycle rack. Patent.

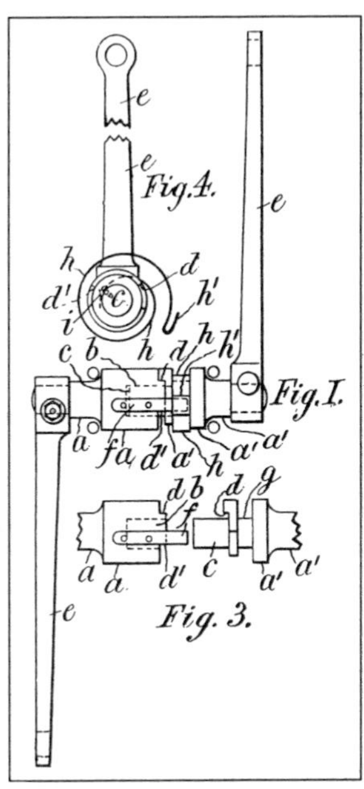

The axle, which enabled both feet to rest in the same position, when not pedalling, devised by Lavinia Laxton. Patent.

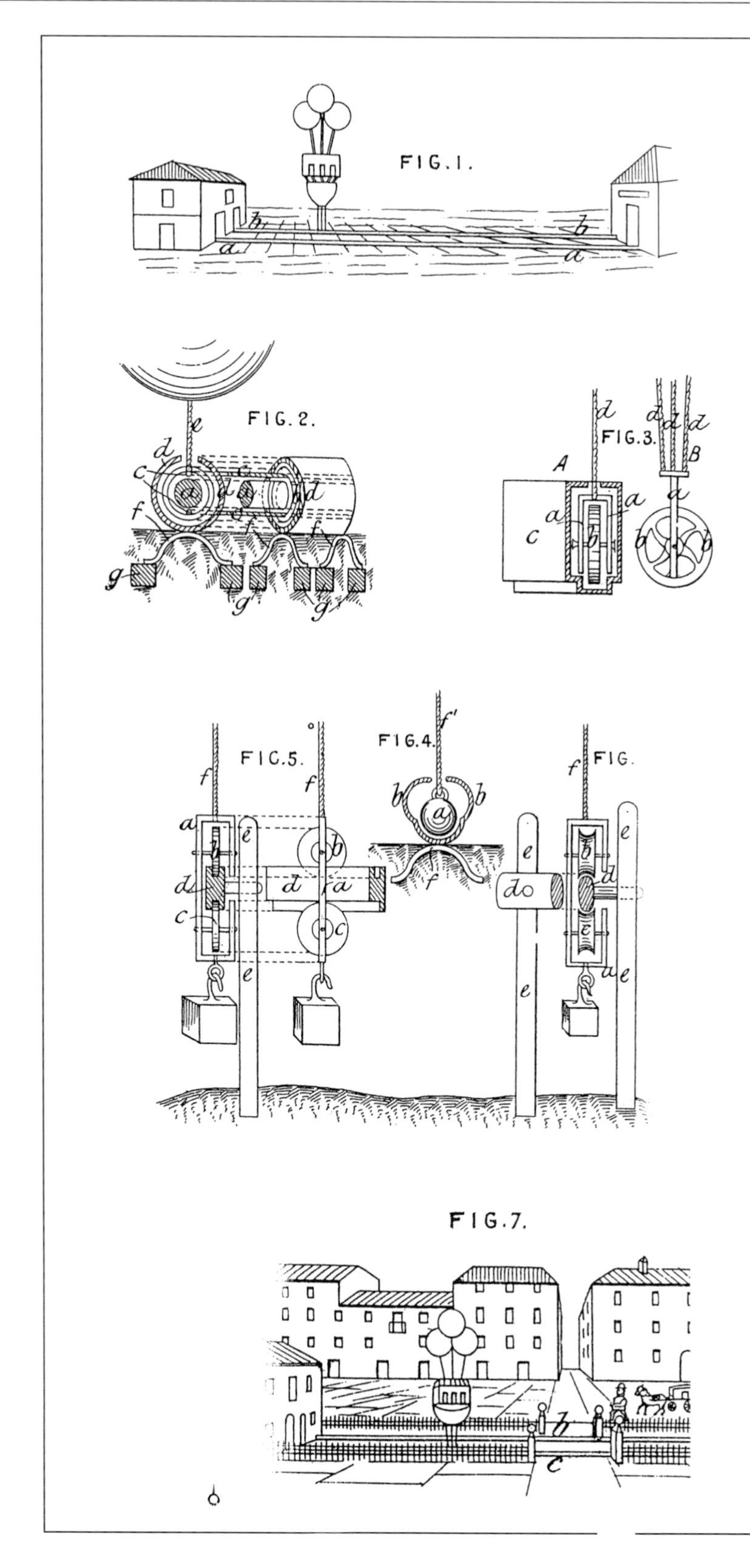

Diagrams for Clara Wells's improvements in aerial locomotion which included a raised railway around the Bay of Naples.

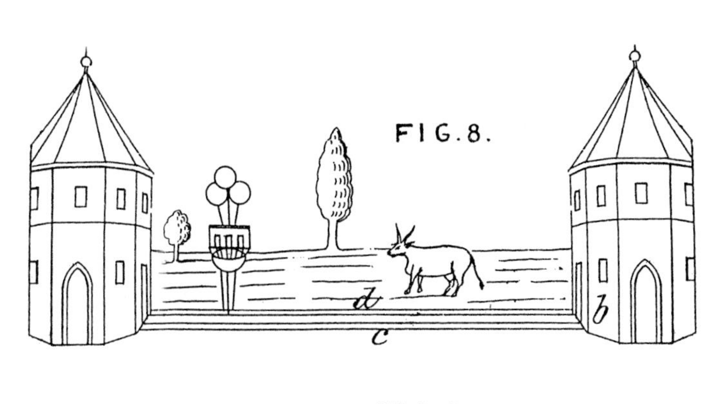
FIG.8.
d
b
c

FIG.9.
c
d
b

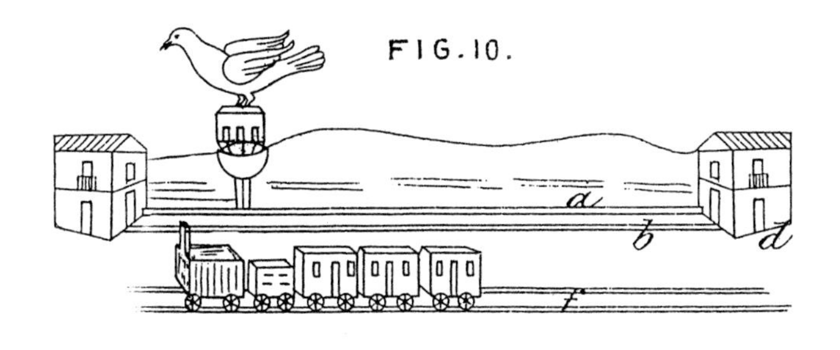
FIG.10.
a
b
d
f

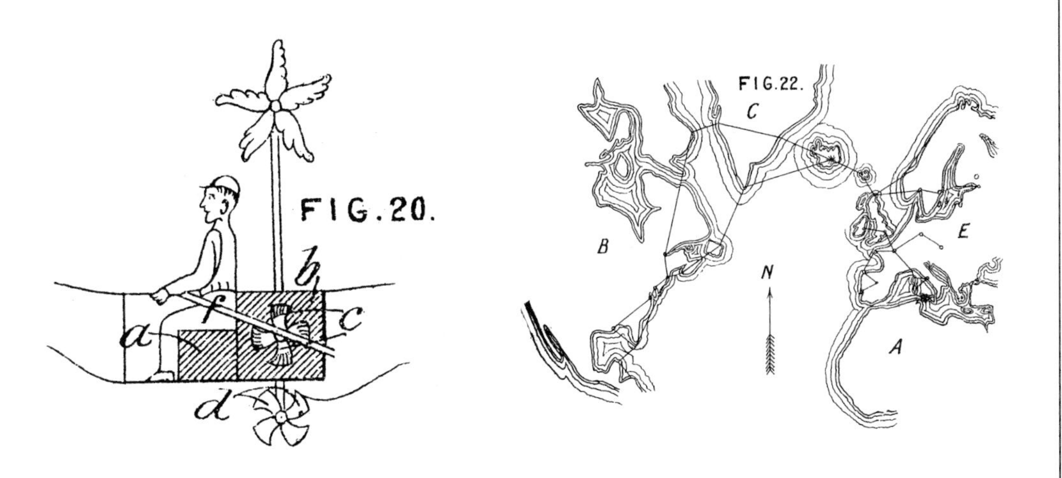
FIG.20.
b
c
f
a
d
FIG.22.
C
B
E
N
A

locations in which she found herself. In a wonderful, free-spirited way typical of many inventors, she put her mind to resolve problems ranging from water storage to aerial locomotion, proposing extensive systems rather than individual inventions. Although her ideas may be regarded as impractical, Clara Wells did not allow inhibition to hinder her imagination as it ran over the boundaries of practicality and custom. The result was a series of complex schemes designed, regardless of budgetary constraints, to tackle what are still pressing problems: supplying fresh water, city and global transportation, and using the meteorological and volcanic forces as power to navigate the world. Maybe she was ahead of her time: with such an imagination and twenty-first century technologies she might have been able to refine her schemes into practical solutions. Instead this boundless enthusiasm has been forgotten.

In 1887, Clara Wells stayed in Naples where the first of her schemes was devised for 'Obtaining Fresh Water from Sea Water for Supplying Towns and for other Purposes'.[32] This was a method of purifying water in which salt water, 'from far enough outside to be perfectly pure', was brought through a pipe and 'boiled in a steam engine, then condensed, or distilled, and boiled, until it becomes good drinking water. Then it must be pumped into water towers erected upon consecutive altitudes' and on into reservoirs from where it would be distributed in pipes.

Maybe for Clara Wells the descriptions of the explorations by Phileas Fogg, the English adventurer created by Jules Verne in his 1873 book *Around the World in Eighty Days*, were influential. Verne himself was fascinated by technological developments and the scientific nature of the earth, which he used in his fantastical novels. Having spent time in Pompeii, by 1890 Clara Wells' schemes were becoming even more ambitious as she combined hot air balloons and overhead railways to conquer the challenges of the natural world. Still living in Naples, she looked at aerial locomotion as a means to resolve pedestrian and traffic congestion in the city. Based on a funicular railway, which she had probably seen in the Alps, Clara Wells devised a series of raised stations linked by an overhead railway system which could be powered by balloons. Even large birds perched on the balloons were included in the scheme. Passengers and their luggage could board at any station and enjoy a ride around the bay.

Her stay in Pompeii, where she saw the results of Mount Vesuvius's drastic eruption, enthused her to find a positive use for volcanic activity in her 'Centres Providing Means for Controlling and Utilizing Volcanic, Aqueous and Meteorological Forces' in 1897.[33] This extensive scheme included a series of isolated 'centres' equipped with appropriate machinery and living accommodation for the workers.

During its dormant period, a series of defensive ditches and trenches would be built around a volcano. These would guide the molten lava away from habitable areas to a depot where it would be left to harden and then transported by sea or rail to a place requiring its use. Heat would be drawn out of the volcano into special receptacles, and then transported through a series of insulated tubes to factories and buildings requiring heat. The tubing was to be laid over a great distance and even underground. Likewise, the gases and minerals would be collected and transported through a different series of pumps and tubes. Those centres near volcanoes would be 'well furnished with convenient means of access, such as railroads, tramways and aerial locomotion, and with hotels and restaurants, as well as various Industrial and curative Establishments'. Hot water could be transported through tubes from geysers and hot springs and high walls built around vineyards and such places to protect them from high winds and cyclones. Fortified 'centres' would also be built for sailors to shelter on isolated, windswept islands to allow them to moor their ships in safety and wait for a tempest or cyclone to pass.

The following year, she combined her ideas for recycling the world's naturally produced energy with her proposal for aerial locomotion to devise a fascinating scheme, to 'secure Modes of Exploring the Cold and Hot Regions of the Earth by means of centers of elevation and depression, with reference to volcanic, aqueous and meteorological forces, and of routes suspended with or without balloons'.[34] Her transportation scheme for the Bay of Naples was now extended to circumnavigate the world. As in her Pompeii-inspired methods, warm air was pumped through a series of valves and large tubes from volcanoes and geysers to colder regions and vice versa, so that explorers travelling over the equator and on to the poles would benefit from it. Travelling on raised railways or in balloons and keeping all lines of communication open between places ensured their safety. In ten lengthy pages she makes her citation in which:

> The Creator of the Universe has established volcanoes and springs, emitting fire and heat, in the vicinity of the North and South poles; and there are very high mountains, their summits covered with snow, in the Torrid Zones and near the Equator.
>
> Instead, therefore, of exposing the lives of courageous men, who endeavour to reach these regions, in frail vessels, steamers, or solitary balloons, hazarding all with little probability of success, this specification describes a secure mode of reaching, either the Poles or the Equator, by using 'centres of elevation and depression, of the volcanic, aqueous and meteorological' forces; combined with suspended and other railways, balloon routes, balloons for raising

> impediments or submerged objects . . . Centres for volcanic forces and hot springs will be constructed . . . advance can be made in whatever direction . . . and by means of large tubes, pneumatic pumps, or the like, carrying the heat along, from one centre to the other, while the necessities for life . . . may be conveyed over suspended railways, balloon routes or the like.

Detailed drawings show the series of stations and suspended tracks and cables that would be constructed across land and ocean. Trains, balloons and other suspended vehicles could carry people and goods around the world powered by steam or electricity. The balloons could be filled with gas or heated and compressed air, which had been pumped in through the large tubing from the appropriate geyser or volcano.

Clara Wells's ideas still seem improbable, but for most of the nineteenth century the widespread use of motor transport would have appeared equally unlikely. Just as the bicycle was revolutionising lives, the prospect of motorised transport to be operated by the people became a possibility. It was while on a motorised tram in 1903 that Mary Anderson came up with a revolutionary idea which would become an essential and legal requirement of all vehicles: 'a simple mechanism . . . for removing snow, rain and sleet from the glass in front of the motorman', the windscreen wiper.[35] She lived in Birmingham, Alabama in the hot climate of the southern United States and chose to visit New York in the snowy winter. Noticing that the tram driver kept getting out to wipe the snow from his windscreen, she saw a problem waiting to be solved. This was a rubber-bladed squeegee on the outside of the window which was fastened to a spindle which went through a hole in the top corner of the window to be attached to a handle on the inside. When the driver turned the handle on the inside, the window was cleared. It could be removed easily when not needed in the summer. It is not known what remuneration, if any, Mary Anderson received for her ingenious invention, which in principle is replicated in millions of vehicles a hundred years later.

Kate Morgan, a corsetière and surgical belt maker trading as 'Kate Wolstenholme' from Wells Street in central London, was possibly influenced by the superior status of her clientèle when she put her mind to the temporary carpeting of motor vehicles and trains.[36] The carpet was on a roller with a spring and weight to keep it flat and taught. A spring plate incorporating a squeegee blade would collect any dirt or mud from the carpet when it was wound back on to the roller. A similar desire for cleanliness inspired Katie Moran from South London to clean the tramlines.[37] She attached revolving scrapers and brushes to the tramcar. They could be raised or lowered

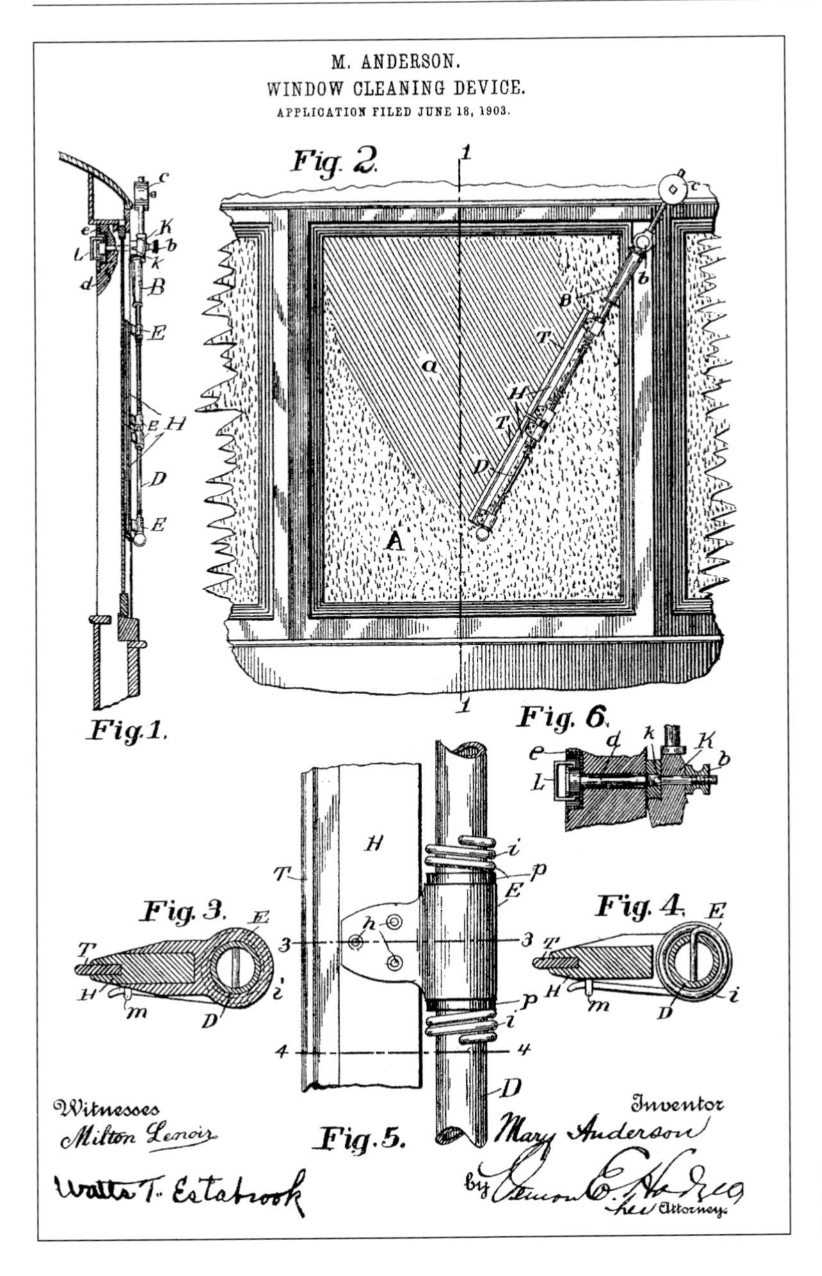

Mary Anderson's windscreen wiper, which she invented for a New York tram. Patent.

with a lever on the tram, and water flowed through tubing from a tank to flush out the dirt.

The arrival of the motorcar brought yet more problems and more scope for women to resolve them. Two women, Sarah Bazeley from Northampton and Annie Birkin of Hanwell, sought to remedy the problem of punctures in pneumatic tyres.[38] When a vehicle was lowered due to a faulty tyre or if overloaded, it would make contact with Sarah Bazeley's gauge[39] attached to the side of the car.

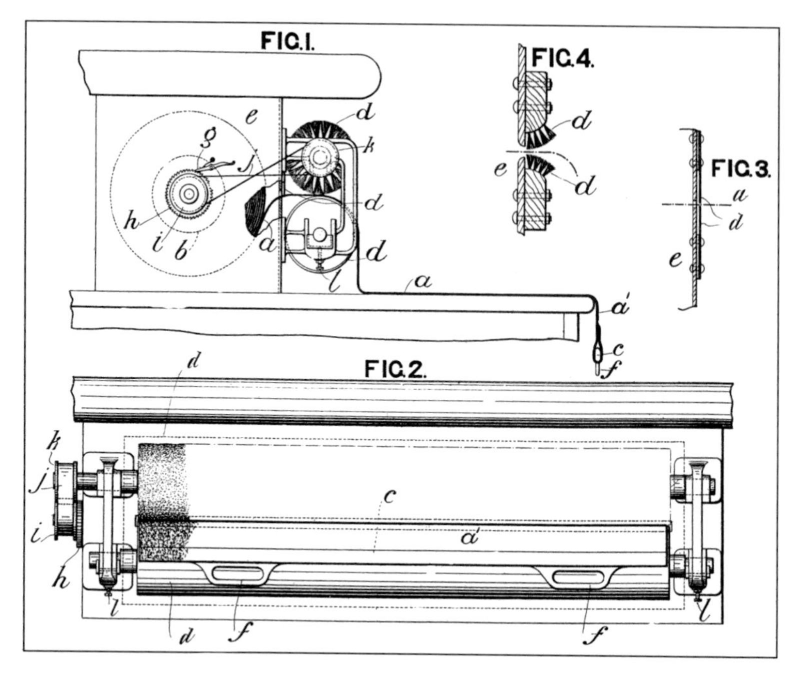

The corsetière Kate Morgan devised this roller from which carpet could be laid in a motor vehicle. Patent.

Mr Rolls in one of his motorcars.
(*Illustrated London News*)

This, being an electric device, emitted a signal to the mechanic and punctures could be detected early, before too much damage had been done to the tyre. Annie Birkin[40] designed an extra U-shaped rim to cover the edges of the tyre giving it greater protection. Justine Lewis-Thiessen from Brussels[41] was concerned with the problem of vehicles skidding and devised a wrapping to go around the tyre made of a double strip of rubber and leather glued together. Holes, lined with metal eyelets, were punched into the rubber and conically

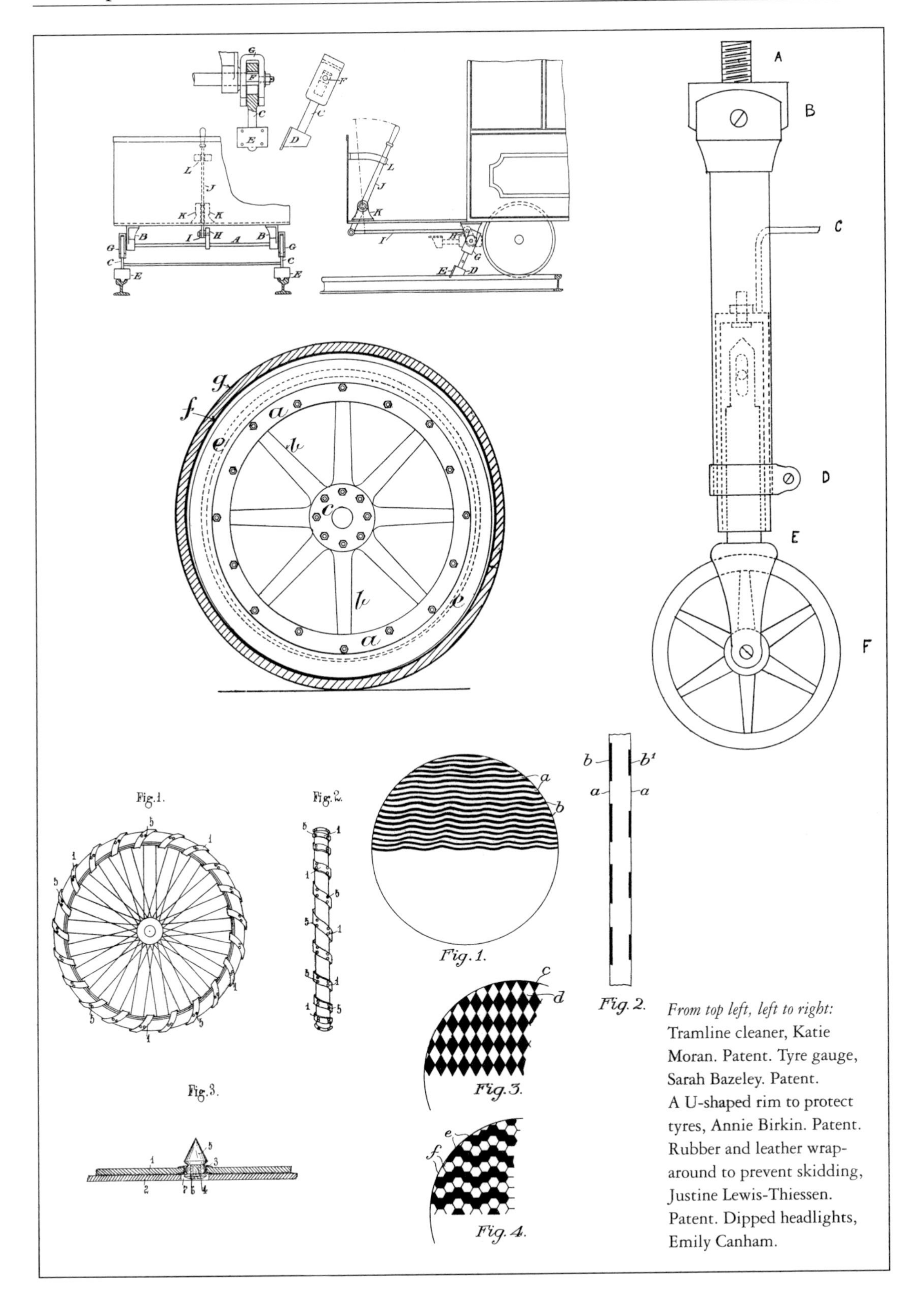

From top left, left to right: Tramline cleaner, Katie Moran. Patent. Tyre gauge, Sarah Bazeley. Patent. A U-shaped rim to protect tyres, Annie Birkin. Patent. Rubber and leather wrap-around to prevent skidding, Justine Lewis-Thiessen. Patent. Dipped headlights, Emily Canham.

shaped steel points screwed into them. The whole combined strip was wrapped around the tyre, while the leather prevented the steel points puncturing the tyre. The points did, however, stop the tyre from skidding.

Dipping the beams of light from car headlights occupied Emily Canham from Highbury, North London.[42] Her proposal to decrease the amount of blinding glare was to divide the lenses into zones and place different opaque or ground glasses or other coloured transparent material over them. The ground glass could be made with patterns of wavy lines or geometric shapes. The bottom zone would be the only one emitting pure, bright light.

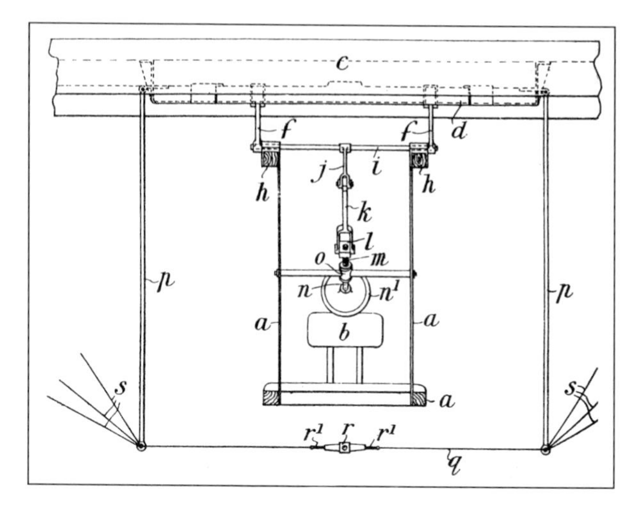

One of Sheila O'Neill's designs to realign the propeller and chassis of a monoplane to improve lift-off. Patent.

Finally, in line with the aspirations to fly, Sheila O'Neill, who described herself as a 'motor car driver', sought to make 'Improvements in Flying Machines' in 1910.[43] Her invention was a realignment of the propeller and chassis so that a monoplane with a front engine could rise from the ground, either automatically or manually, and gain greater speed to lift the plane. She was also able to move it backwards, into a gliding position, should the engine fail and ensure a safe descent. Sheila O'Neill found that

> by mounting the propeller in such a position within the frame chassis, or fuselage tressle, so that it pulls slightly in an upward direction, there is less necessity to have the plane set at a deep angle . . . thereby saving the loss of power usually caused by skin friction and air resistance from atmospheric pressure.

The complex detail of her technology with its interdependence of rods, hinges, tressles and turntables could all be folded for transit by road or rail. Sheila O'Neill was, no doubt, an explorer of technology and as enthusiastic for the new aeronautics as she was for the motor car.

CHAPTER EIGHT

Queens of Science and Medicine

Even though their contribution to science and medicine is still regarded as having been minimal, women were, from the eighteenth century onwards, involved in scientific debate and many have made significant discoveries, implemented changes and devised inventions that have affected the lives of generations. However, their access to the science laboratories and into medical schools was usually blocked and, as in the suffrage movement itself, many fought long and hard battles for acceptance.

In her fascinating study, *The Scientific Lady*, Patricia Phillips describes the scientific world and women's access and contribution to it, as well as the snubs and disdain they received from the male hierarchy. From its inception in 1799 in Albemarle Street, the Royal Institution welcomed women as members of its audiences. To attend such lectures was part of the social scene and further emulated throughout the country with the foundation, in 1831, of the British Association for the Advancement of Science and at local scientific and literary institutes. So women were very aware of the latest scientific discoveries, inventions and debates but continued to be denied access to committees.

It is likely they would have watched Michael Faraday's demonstrations to show the relationship between electricity and magnetism at the Royal Institution in the 1830s. The use of electricity for power was embraced more in continental Europe than in Britain where the power base of the Industrial Revolution was already established on steam.

One female enthusiast, Mary Somerville, was known as the 'Queen of Nineteenth-Century Science'.[1] Others were mathematicians, botanists, geologists, physicists and later doctors and engineers. Mary Somerville was a friend of Augusta Ada, Countess of Lovelace, the daughter of Lord Byron, and introduced her to Charles Babbage. Ada Lovelace's mathematical knowledge enabled her, in the 1840s, to work with Babbage on his analytical engine which, it is claimed, was the first computer. She was able to work out languages for the machine and was essential to the success and importance of Babbage's invention. One hundred and thirty years later, the ADA computer programming language was named after Ada Lovelace.

Augusta Ada King, Countess of Lovelace, etching by Alfred-Edward Chalon, 1852. (*The Royal Institution, London/ Bridgeman Art Library*)

Sarah Johnson and her husband Alfred used mathematics to aid builders, no doubt after watching them interpret architects' plans for the 1870s building boom. Their 'Improved Plumb Rule and Level'[2] was to enable builders to mark horizontal and vertical lines and a line at any angle between the two. It was made of a wooden level with a metal dial in the middle. A metal weight was suspended from a needle on the dial so that the needle always pointed vertically. As the level was moved around, the dial would move but the needle stayed in the same position, thus ascertaining the angle. Sarah Marks (Ayrton) designed a mathematical divider in 1884[3] to divide a line easily into set lengths. A friend of George Eliot she was married to William Ayrton, a professor of physics, who supported her in her scientific interests, and although she was a member of the Institution of Electrical Engineers, the Royal Society refused her admission because she was married. She went on to design arc lamps and flapper fans which were used to clear the trenches of noxious gases during the First World War.[4] Lillian Sangster from Redhill in Surrey devised her 'Improvements in Ruling Attachments for

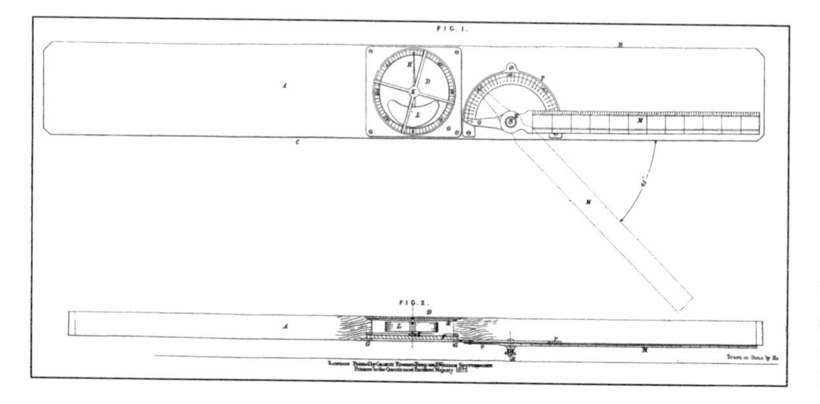

The 'Improved Plumb Rule and Level' for use by builders designed by Sarah Johnson and her husband. Patent.

Measures of Length, being Two Instruments to Fix on Measures, to Measure and Mark at the same Time'.[5] Her device could be made to work with ink, pencil or crayon and was intended for use in schools and offices, by carpenters and dressmakers.

The use of scientific knowledge for a technological purpose and in industry increased towards the end of the nineteenth century and a number of women explored the implications. Since 1891, the Royal Commissioners for the Exhibition of 1851 had used the profits from the Great Exhibition to fund research scientists and, between 1891 and 1900, 10 women and 152 men received awards.[6] Elizabeth Barnston Parnell, an inventor and metallurgist, took out two patents in 1889 and another in 1891 for methods to extract metal from ore and for improvements in calcining furnaces. She was an exhibitor at the Chicago Exhibition in 1893.

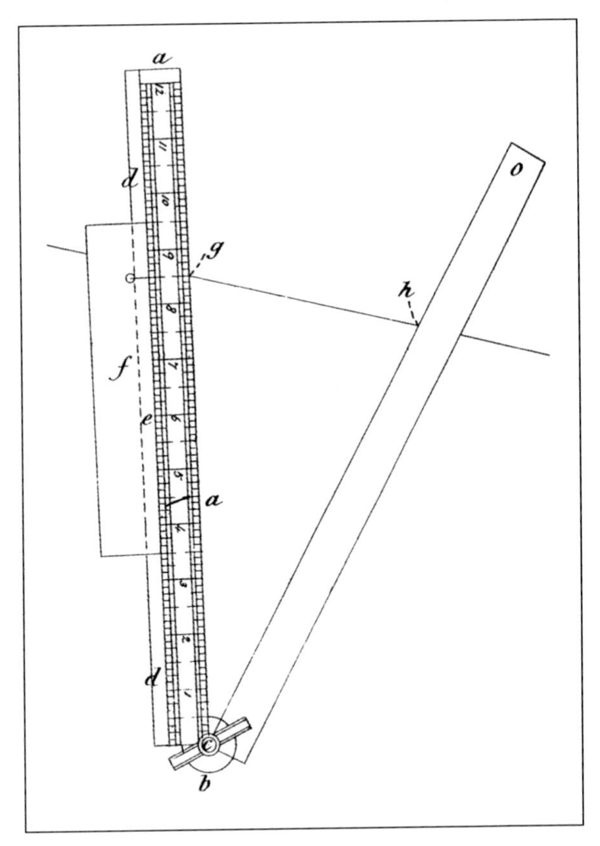

Above: The divider to dissect a line into set lengths invented by Sarah Marks (Ayrton). Patent.

Mary Hill and her husband John, both jewellers and gas fitters from Trimdon Colliery in the Durham coalfield, devised 'Acetylene Gas Generating Apparatus'.[7] Their series of diagrams describes how by automatically regulating the water supply to calcium carbide the generation of the gas would be improved.

Below: Schoolchildren, office workers, carpenters and dressmakers could all use Lillian Sangster's rule and measure. Patent.

Although some women were successful in transgressing the gender boundaries in the field of science and contributing to its understanding and use, for the majority, involvement in it tended to rest with more traditional roles. As wives and mothers they had run households, nursed their families when sick and tended to all their needs. Poultices had been applied, tinctures and medicines administered, labouring women attended to, newborn babies nursed and the dying comforted. With the improvements in sewerage and sanitation, resulting from Edwin Chadwick's 1842 report, and a better understanding of the

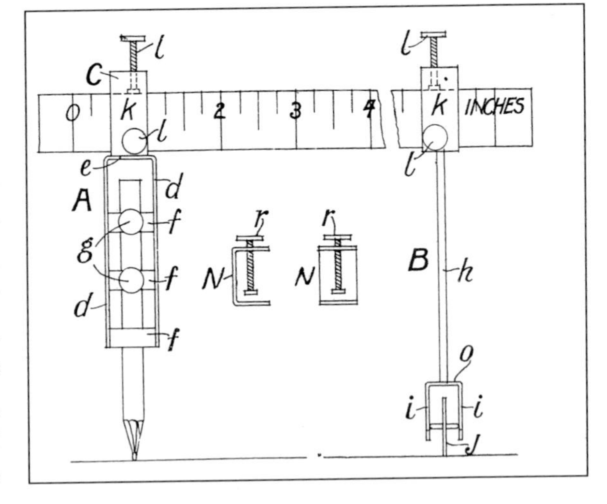

spread of disease and basic hygiene, epidemics of cholera, typhoid and other deadly diseases had been reduced. But tuberculosis continued to threaten families. The medical profession, too, was changing, as doctors like Thomas Wakley, the founder of *The Lancet*, demanded it be more professional and questioned the role of the quack. But for women, again, entry into the medical schools to train to be doctors was a battle to be won. Elizabeth Blackwell became the first woman doctor in the USA in 1849. Elizabeth Garrett Anderson, the first woman doctor in England, had to study medicine privately because she was not allowed admission to medical school. However, in 1866 she became a medical attendant at St Mary's Dispensary, London which was renamed the New Hospital for Women and at which she was able to establish a medical school for women. This hospital became the Elizabeth Garrett Anderson Hospital in 1918, a year after her death. Sophia Jex-Blake, having achieved a place on a medical course at Edinburgh University, was not allowed to receive a degree, but fought to change the legislation to enable women to be justly rewarded, and she won in 1876. She opened the first medical school for women in London in 1874 and another in Edinburgh in 1886. Even Queen Victoria's actions astounded the medical professions when she irretrievably changed the course of childbirth for future generations of women. She insisted on being administered the new chloroform, as an analgesic against the pain, for the births of her last two children, Prince Leopold and Princess Beatrice. Also, on her insistence, a midwife was in attendance, alongside the doctors, during the births of all nine of her children.

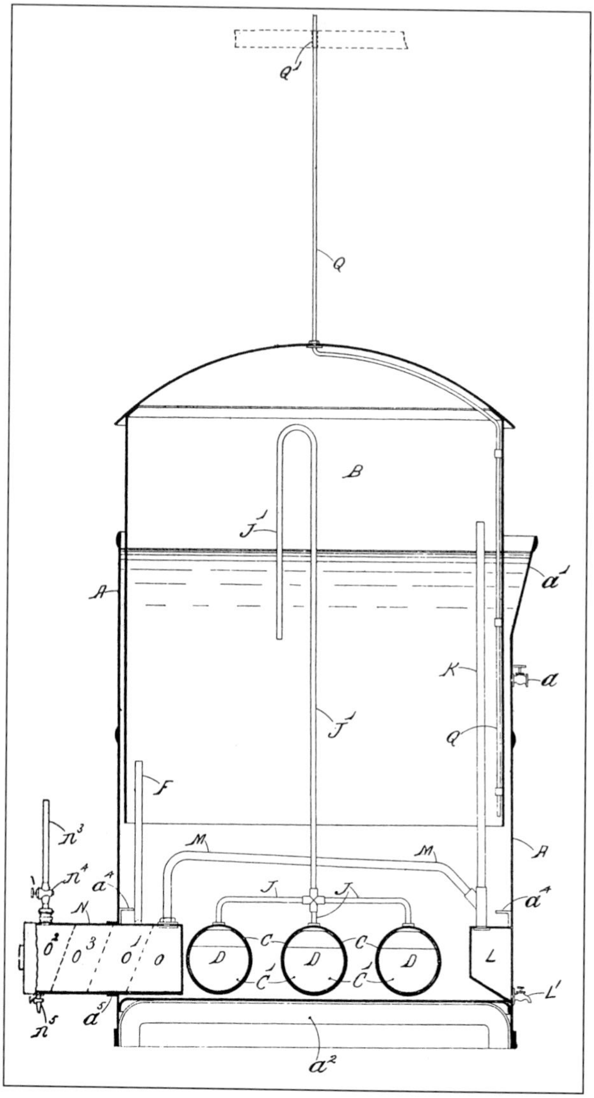

Part of the 'Acetylene Gas Generating Apparatus' by Mary Hill and her husband John. Patent.

It is easy, with hindsight, to look at the diseases endemic in earlier centuries and be appalled at the methods of dealing with them used at the time. It was not then known that rickets and scurvy are caused by vitamin deficiencies; that cholera and typhoid are the consequence of water supplies contaminated by sewage. Many families were ravaged by tuberculosis, which spread rapidly in poor housing conditions. Intestinal worms were prevalent, so Rebecca Ching's lozenges, designed to rid the body of them, were not unusual. The widow of John Ching, she improved on his method in 1808.[8] Made of 'sub murias hydrargyri, 1lb; saccharum album, xv lbs; crocus sativus 3fs; aquae fontanae, 1lb' which was boiled, mixed and strained then rolled out using a brass roller, to a thickness of ⅛ inch. Oval lozenges were then punched out of the yellow substance with a tinplated cutter. Mrs Ching recommended taking between one and six lozenges each night and another brown one, made with an extract of convolvulus, in the mornings. A hundred years later, Ellenor West from Guildford advocated an ointment made from butter, elder rind, silgreen and potato to treat burns, eczema and bleeding piles.[9]

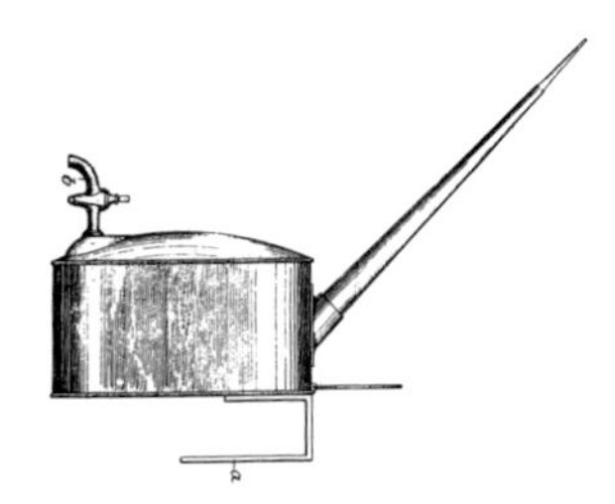

Emma Pike's Bronchitis Kettle was clipped on to the front of the cooking range and vapours escaped through the funnel to fill the room. Patent.

Tuberculosis is highly infectious and its effects are still devastating. Among noted figures, the poet John Keats and Ann and Emily Brontë succumbed to it. There were numerous recommendations and devices to attempt to cure it. Elizabeth Spink from York proposed her 'Liniment for Lung Disease'[10] in which she made a paste from powdered nutmeg, nutgall and calomel mixed with wine. Once the wine vaporised, she mixed the powder with fat to rub into the body as an embrocation or liniment. This, she claimed, would, if rubbed into the chest, back and shoulders of 'persons suffering from consumption, phthisis, and other lung diseases', cure (or relieve) such complaints.[11] As calomel is made from mercurous chloride it is possible that the patient would have been poisoned rather than cured of their primary ailment. The symptoms of lung and chest complaints could also be relieved by Emma Pike's 'Bronchitis Kettle'.[12] To avoid the inconvenience and dangers of removing a boiling kettle from the range to create a moist atmosphere for bronchitis sufferers, she clipped it on to the front of the range and the vapours flowed through a funnel on its top and filled the room.

The curative possibilities of electricity were explored. In the 1860s there were many experiments using electricity for medical benefits,[13] including Marie Joséphine Elisabeth Jullienne's 'Bath Belt to be Applied in the Bathing Vessels, and in Electrical Apparatus Connected herein'.[14] This was a French invention to support 'in a bath in the most agreeable manner and without least fatigue, either children of all ages or sick adults'. This iron or steel belt was coated with paint, varnish or a metal that would not rust in the water, covered in horsehair and serge and strapped to the body. A back piece, with a pillow, was riveted to it and screwed to the side of the bath. On the children's version a board was added on which to place toys. Then the electricity was applied:

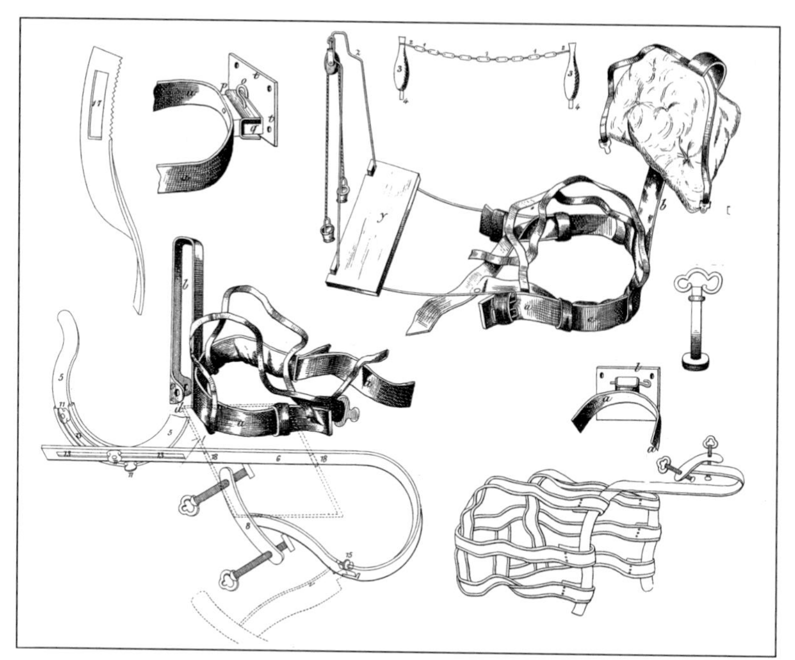

Young children and sick adults were supported in the bath with Marie Joséphine Julienne's belt. Electric shocks, for curative purposes, could be transmitted through it. Patent.

> The apparatus is further arranged so that electric shocks may be communicated to the bather in certain proportions of static or dynamic electricity or magnetism. I have devised a certain arrangement of electric chain, when the apparatus is intended for conveying electricity that will produce the dynamic fluid.

By the mid-nineteenth century the first professional nurses appeared. The onset of the Crimean War excited the interests of many young middle- and upper-class women who, having rejected their families' determination to see them respectably married off, and probably with philanthropic and romantic idealism, wanted to go and help nurse the sick. Florence Nightingale and Mary Seacole were just two of many women who sailed off to the Turkish/Russian borders to find hordes of injured, sick and dying men lying in cramped hospital conditions. They and the doctors quickly became aware that more men were dying of diseases contracted in the hospitals than of their injuries on the battlefields. Again Edwin Chadwick's 1842 report was fundamental to changing practices: this time in the hospital ward. The concept of fresh air and ventilation to reduce the spread of disease was introduced. Florence Nightingale, Mary Seacole and the other nurses radically altered hospital design to create the long wards, with numerous windows and beds well spaced apart. The horrendous conditions that they witnessed and the deaths of so many young men, determined Florence Nightingale to change hospital practices on her return to England. She informed Queen Victoria of the appalling situation and persuaded her to visit the

Above: Florence Nightingale tending the sick, injured and dying at the hospital in Scutari during the Crimean War. (*Illustrated London News*)

Below: An Operation at Charing Cross Hospital. (*Private Collection/Bridgeman Art Library*)

wounded soldiers. Queen Victoria did as she was asked and travelled to Chatham Hospital to meet the survivors of the Crimea. Florence Nightingale could hardly have asked for a more prestigious ally than the Queen and, in 1854, her demands for a Royal Commission on Hospital Administration were met. Florence Nightingale has become an iconic figure in the history of nursing but there were others, notably Mary Seacole who was born in Jamaica. In the Crimea she established the British Hotel for wounded soldiers. At the end of the war and without money, her supporters in England launched a campaign in *The Times* to bring her to England, where she died in 1881 leaving a legacy for Black women to become nurses. This still continues at the Mary Seacole Centre for Nursing Practice in West London. Florence Nightingale had become an expert in the management of hospital wards and was instrumental in the design of the new St Thomas's Hospital when it moved from Southwark. Originally, based

on her belief in the benefits of fresh air, she had wanted it to move to the country, but settled on the current site by the River Thames in Lambeth, opposite the Houses of Parliament. Here she opened the first nurses' training school.[15]

The expansion of nursing as a profession is reflected in the number of inventions registered by nurses and midwives at the end of the nineteenth and early twentieth centuries. Most were of quite specific items, clearly drawn from the requirements of their work. Sarah Norledge, the matron of a nursing home in Boscombe, Hampshire, devised an 'Extension Apparatus for Surgical Purposes'.[16] This consisted of a frame, fastened to the foot of the bed, attached to which was a tube 'through which runs a cord having at one end means for attaching it to the leg of the patient, and at the other end a weight, a pulley being provided for the cord to pass over'. Cross bars and rigid hooks maintained the frame's stability and 'absence from vibration'. Florence Hurd of Highgate, Middlesex made 'Improvements in Appliances for Inducing Correct Breathing through the Nose instead of the Open Mouth; for Flattening Prominent Ears and Keeping Surgical Bandages and the like in place on the Head'.[17] Another method to keep bandages in place was developed by a trained nurse, Beatrice Kent,[18] from London. Before the invention of adhesive dressings, bandaging wounds could lead to many unwanted side effects:

> With certain operations to the lower parts of the body great difficulty has been experienced in bandaging the patient in such a way that the dressings shall not slip, especially when the patient moves about . . . with some operations the risk of accident to the patient through the dressings becoming disturbed has been so great that it has been necessary to keep the patient in bed for a longer time than would otherwise be desirable.

Beatrice Kent's solution was to make a wide body belt out of washable material that was fastened with buckles around the abdomen and shoulder straps, to keep it in place over the bandages. Additional belts could be added to cover the thighs if necessary.

One of the detrimental effects of tight stays and corseting was the effect on the nipples of nursing mothers. The intense pressure of the corset on the nipple caused them to become inverted making it very difficult for a newborn baby to suckle. From the seventeenth century onwards nipple shields, to place over the inverted nipple, were made from silver, wood, leather and glass (or hollowed-out walnut shells for poor women) to replicate the shape of the perfect nipple. Then the baby sucked the milk from the breast through the

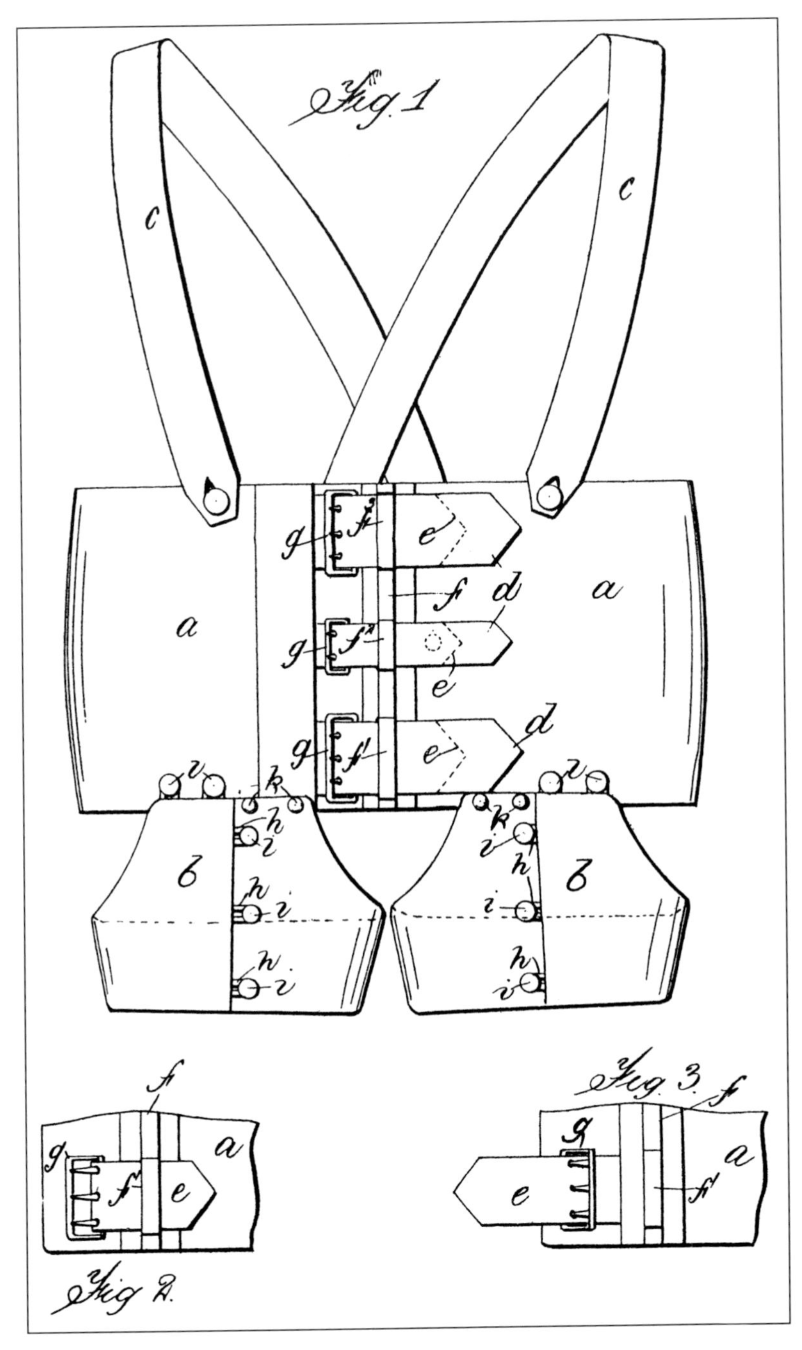

Beatrice Kent's washable body belt to keep surgical dressings in place. Patent.

holes drilled in the shield. By the 1850s nipple shields were being manufactured, some made from India rubber.[19] Using her experience as a registered medical, surgical and maternity nurse, Edith Whittaker, from Cheshire,[20] in 1905 outlined the problems of existing nipple shields and the problems encountered by newborn babies unable to suckle:

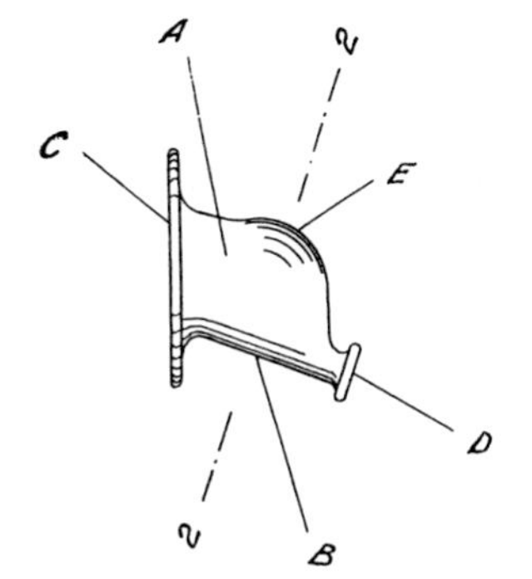

A glass nipple shield with an India rubber teat designed by Edith Whittaker. Patent.

> An objection appertaining to the present shape of nipple shield, which consists of a glass shield provided with an India rubber teat, is that considerable suction is requisite before any milk reaches the mouth of the infant. This objection is due to the present shape of the glass shield, its bell-like contour not readily conducting the milk into the teat . . . the infant should quickly taste the milk, so that he will be induced to continue sucking; unless he has this encouragement he is apt to discontinue his efforts.

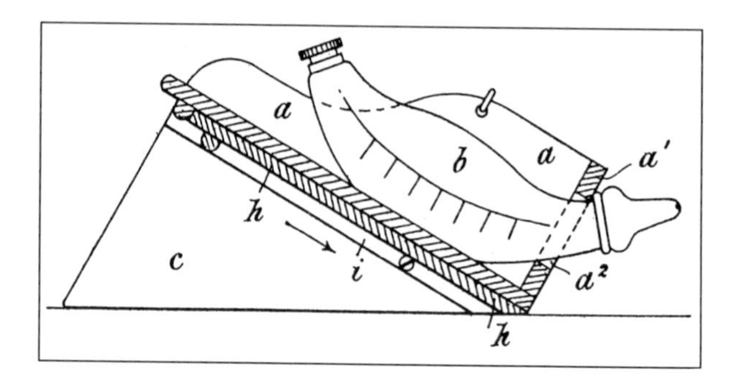

The bottle stand enabled Kate Bargery to feed more than one baby at a time. Patent.

Her proposal was to make a shield, the bottom of which formed a channel so that the milk would pass directly from the nipple through it into the teat and the baby's mouth. The bulging upper part of the shield formed a reservoir for any excess milk and the shield would be turned upside down for it to be sucked away by the baby. She also made the shield sit at an angle from the nipple, pointing downwards in line with the baby's mouth rather than horizontally as in the past. Another nurse, who worked at the Mildmay Hospital, Bethnal Green, in London's East End, was Kate Bargery. Her problem was how to bottle-feed 'a large number of infants (who) have to be fed at one time as in nurseries, hospitals and other places'. This problem she resolved by making a wooden bottle stand, from which her babies could suckle without having to be held.[21] The sterilising of instruments was beginning to be understood and Alexandra Richter from Hospitalstrasse in Leipzig, Germany invented an 'Improved Sterilising Apparatus'.[22]

An early hospital steriliser invented by Alexandra Richter. Patent.

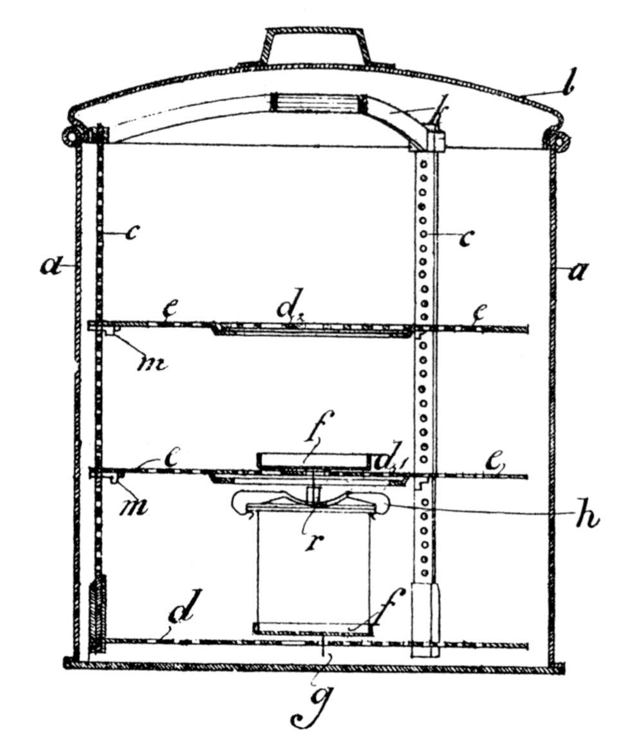

During the nineteenth century the new concept of the 'industrial accident' appeared.[23] In factories there was scant regard for safety – small children were employed to operate machines and even climb inside to maintain them – and there were terrible disasters on the new railways. Stoked by fires, engines would ignite and engulf the carriages, and trains might derail off bridges or down embankments.

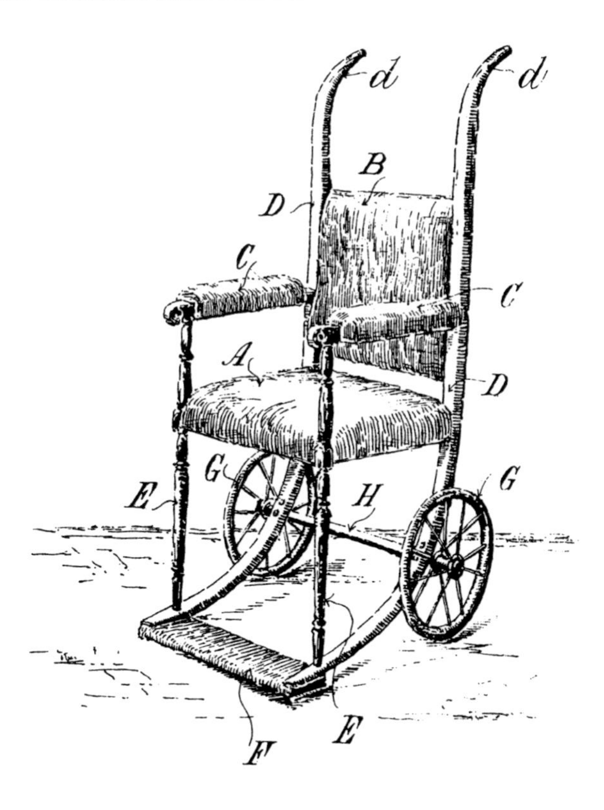

The patent diagram for Fanny Smith and Astley Carrington Roberts's wheelchair. Patent.

The victims of these accidents, as well as the hundreds of soldiers returning wounded from the battlefields abroad, required attention. Inevitably, there were inventions to meet the needs of the injured and disabled. In 1853, Mary Davy from Homerton and Ann Taylor of Islington focused on the bathing needs of people whose arms were weak or had been amputated. Their 'Mechanical Application of Brushes' was a complex arrangement of metal plates, a working barrel, springs, axles and drums to make a brushing machine for the bath, to enable the person to bathe in privacy without supervision.[24] Later, in 1862, Emma Duncan of Bayswater, possibly having seen something similar on travels in the Far East, made splints from natural materials[25] not indigenous to Britain: 'rattan, bamboo, or jungle cane, either split or entire, and in longitudinal rows or layers', making them much lighter and cheaper than the traditional heavy metal. Wool or cotton wadding padded the hard frame. Her arm slings were made of woven cane with only metal at the base so that they could fit an individual's shape.

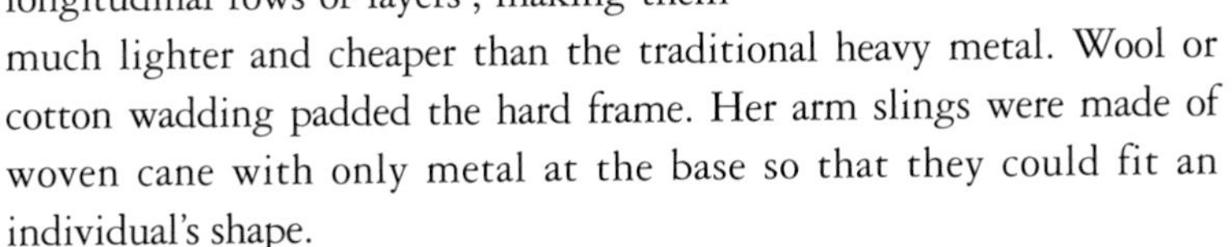

Emma Smith of Weston-super-Mare designed a multifunctional writing desk for the blind, specifically to enable them to write by hand.[26] The first wheelchair had been designed by James Heath from Bath in the 1750s. This was more like a carriage than a chair on wheels, with its three wheels and hood, and could be pulled by a person or horse. During the nineteenth century they were used not only for those with disabilities but also for pleasure rides. Such vehicles were popular at the new Victorian seaside resorts like Blackpool and Southend. Often equipment for the disabled had been poorly designed and cumbersome, drawing attention to the user rather than trying to merge with the environment. In contrast, in 1899 a spinster, Fanny Smith, and a surgeon, Astley Carrington Roberts, both from Eastbourne on the Sussex coast, designed a very elegant 'Invalid Chair'.[27] Clearly influenced by the Arts and Crafts movement, this was a simple, wooden, high-backed armchair with an axle and wheels instead of legs at the back. The extended side rails were curved, the front legs and foot-board resting on them. They then curved along the sides and up the back where they formed

A recently discovered Smith-Roberts chair. (*Private Collection*)

elegant handles. An example of this chair has recently been found in a loft in the Midlands. While Dr Carrington Roberts appears on the 1901 census as a 38-year-old medical practitioner, it could be conjectured that Fanny Smith was an elderly patient and they worked on this design together.

The concept of a portable hospital for army use was explored by Isambard Kingdom Brunel, who designed a wooden, prefabricated hospital to encompass the new nursing procedures required for the Crimea. However, the war had ended before it was put into use. Emily Mitchell from Torquay probably followed Brunel's thoughts with her 'Improvements in and relating to the Walls of Tents, Huts and the like' in 1903.[28] Hers was a large tent in which the flow of fresh air could be regulated and draughts controlled making it suitable for the treatment of patients with tuberculosis and for use as a field hospital. Within, it had a series of canvas walls to make cubicles that could be left open or closed. The curtains or walls were made of fabric hung from metal rods at different heights. Madame Roels, from Brussels designed a stretcher, made of a frame with strapping across it, on which a patient could be carried and supported in different positions.[29]

The problems of dealing with the ablutions of the bed-ridden, without using an uncomfortable bedpan, were faced by Ada Dugan from Londonderry and her 'Improved Mattress for Invalids'.[30] This mattress was especially suitable for those with paralysis and other weaknesses; its upper part was made of horsehair and the lower of wire. A hole was

Emily Mitchell's portable tent hospital. Patent.

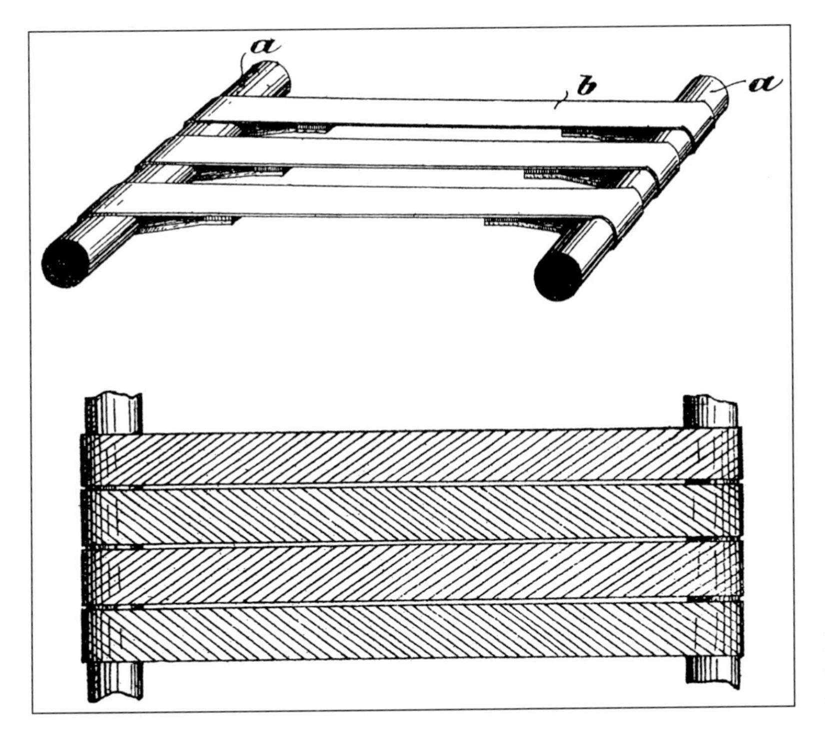

Patients were supported and carried on Madame Roel's stretcher. Patent.

made halfway down, going through both parts of the mattress.

> A waterproof sheet is laid on the upper mattress and is provided with a tube or funnel which passes through the holes in both mattresses so as to deliver to a slop bucket or other vessel placed below the bed . . . the holes in the hair mattress can be lined with any waterproof material . . . and when the pail is removed a pad or pillow can be placed over the hole to prevent a draught.

In the same spirit of helpfulness, a Dublin nurse, Isabella O'Connor, made knitted, woollen covers to put over hospital urinal vessels to prevent the user from being chilled.[31] Hygiene was the major consideration in the 'Sanitary Metal Commode' designed by Mary Dunstone and Emma Bartlett from Brixham in Devon.[32] Their commode was to be made from metal, and not wood as was common at this time. The seat and all the internal parts could be easily and quickly removed for cleaning. 'In view of this construction and the fact that the whole is made of metal in lieu of wood as is customary, the harbouring of microbes and dirt is prevented and all risk of unpleasant odours avoided.'

The douche as a method of contraception became popular, although cumbersome to use. Bertha Chmielewsky (or Sherwinter), a married woman from Edinburgh, addressed the problem by contriving a backrest with which women could douche themselves more easily

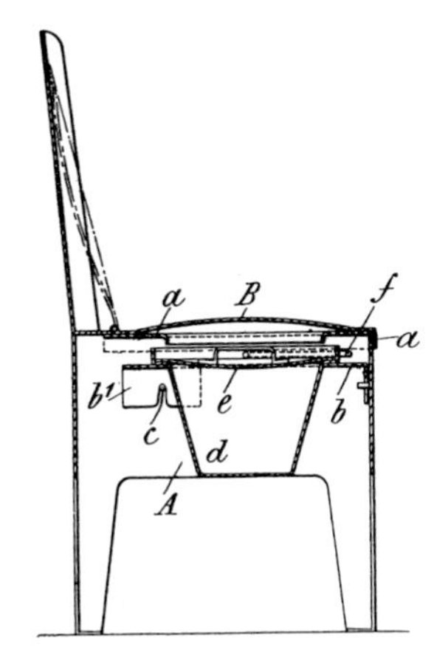

The 'Sanitary Metal Commode' devised by Mary Dunstone and Emma Bartlett. Patent.

than using a douche bath, which was uncomfortable and caused backache.[33] Her back rest was a wooden frame covered in calico to be used in the bath or bed

> to enable women to perform the operation of douching their private parts and adjacent internal organs with greater facility than is attainable by means of the douche-bath at present in use. The discomfort and strain to the back . . . are obviated by using the back rest or bed . . . The hips of the user rest on the bottom end of the frame of the back rest or bed while her legs are stretched out so that her feet may rest upon the sides of the bath.

As in the diagram this complicated and potentially messy manoeuvre could be done while fully clothed: '[this] ensures a maximum of ease and comfort and permits of the easy introduction of the douche pipe into the vagina, and for the water or other liquid employed . . . landing, as it returns, into the bath, and clear of the user's clothes and person.' Elizabeth Keswick from Ilkley was also concerned with administering the douche or enema. She invented an 'Apparatus for use in connection with Enemas, or Irrigators, and the like Injectors'.[34] She describes her apparatus as a metal stand or pedestal in two parts with a seat upon which the person sat. The lower part contained water to go into the douche and the upper the waste from it.

One woman's enquiring mind led to major discoveries which continue to have far-reaching effects around the world. Marie Curie was born Marie Sklodowska in Warsaw in 1867, but left to study physics at the Sorbonne in Paris in 1891 where she met her future husband Pierre Curie. They spent their honeymoon cycling around France. While studying uranium compounds she found that pitchblende was more active than uranium. Together the Curies pursued their search for a new radioactive element and in July 1898 they named their discovery polonium after Poland. They continued their research and discovered radium, which was 900 times more radioactive than uranium. Marie managed to burn herself with some of the radium phials, which led to

Above: This backrest, designed by Bertha Chmielewsky, supported women while douching.

Below: Elizabeth Keswick's douche on a pedestal seat, with a reservoir beneath. Patent.

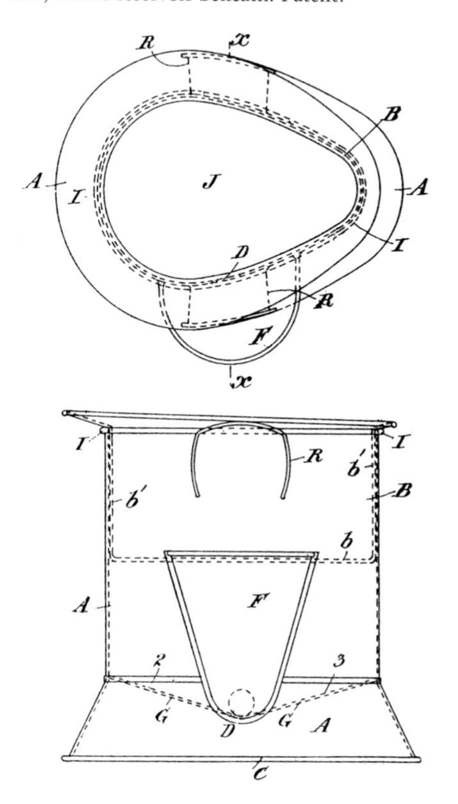

her looking at the effects of it on the skin. In 1903 she found that radium rays could destroy cells; this marked the beginning of the use of radiation in the treatment of cancer cells and other diseases, which is still used today and for which discovery she was awarded, with her husband Pierre and Antoine Becquerel, the Nobel Prize for Physics. When Pierre Curie was killed in a road accident in 1906, Marie became the first female professor at the Sorbonne, was awarded her second Nobel Prize in 1911 and later set up the Radium Institute.[35]

Marie and Pierre Curie in her laboratory. (*ACJC – Curie and Joliot Curie Fund*)

In 1868 a group of three men and one woman, Mary Louisa Booth, all from New York, took out a British patent for a complex method to embalm a body.[36] The description is lengthy and rather gruesome but gives great insight into a process which one might never otherwise have examined. The body was washed in acid and water, had acid poured into slits made in it and was laid on an antiseptic cloth. Carbolic acid and sawdust were placed in the bottom of the coffin before the body was placed inside it.

CHAPTER NINE

Exhibitions: London and America

The phenomenon of the international exhibition, to promote innovation, design, industry and trade, originated in London. During the 1840s the Society for the Encouragement of Arts, Manufactures and Commerce (Society of Arts) held a series of successful exhibitions to promote British inventions and industries at its house in John Adam Street in London. The Society of Arts encouraged women – and they exhibited inventions and designs – but in line with other societies of the time they were not allowed to sit on any committee. Women were, however, recipients of the Society's medals, for example, Jemima Goode in 1843, whose award was for her roller blind for a gothic window. The Society was also instrumental in supporting the straw plaiters in Luton when there was a slump in trade owing to the lifting of the trade blockades following the Napoleonic Wars.[1]

Wishing to improve on the success of their exhibitions, Henry Cole, and other members of the Society's Exhibition Committee, visited the national exhibitions that were being held in Paris and Brussels. These were intended to promote and sell local products and any idea of including foreign goods was, at that time, inconceivable. In 1848, following these visits, and in line with the Society's aim to promote British manufactures and commerce, Cole and his colleagues considered the possibility of making their exhibition bigger than any previous one by inviting foreign traders to exhibit. From the outset, Queen Victoria's husband, Prince Albert, supported the aims of a steering committee from the Society of Arts, even altering the minutes in his own hand,[2] to make this a 'Great Exhibition of the Works of Industry and of all Nations': the first international exhibition of manufactured goods in the world.

With breathtaking speed and within two years of the decision to hold the exhibition, committees were formed, a site was found in

The Grand Entrance to the Great Exhibition, from the official Catalogue of the Great Exhibition, 1851. (*Author's Collection*)

Hyde Park, design competitions held for the building (in which none of the original entries was deemed suitable by the planning committee), and Joseph Paxton's revolutionary, prefabricated iron and glass building was eventually commissioned and erected, to be filled with exhibits from all over the world. When the Great Exhibition opened in May 1851, it was much more than a shop window for British industry. There was a mixture of pride in Britain's manufacturers, who were deemed the strongest in the world, and a desire to improve upon it by provoking international competition. Britain was in a very strong position; its colonies and territories did, inevitably, cover much of the globe. Paxton's building was highly innovative and is recognised as being one of the most important influences in the development of the Modern Movement in architecture. People attended the Exhibition from all over the world; Thomas Cook offered an all-inclusive excursion ticket – return rail fare, overnight stay in London and entrance fee – and brought millions to London for the first time. Rich and poor alike stared in amazement, bewildered and excited at this great palace of crystal; the displays inside were no less exciting as they saw objects and materials previously unimaginable. This experience of novelty, in an era before mass travel, television and virtual reality, must not be underestimated. In comparison, the experience of visiting modern exhibitions and fairs is often a disappointment as so little is truly new. A group of women straw-hat makers from Luton hired a charabanc and made a dramatic entrance. In 1851, for 1*d*, people

First page of the three-volumed *Official Catalogue of the Great Exhibition, 1851*. (*Author's Collection*)

used the first public toilets. They drank bottles of Schweppes mineral water, ate one of thousands of Bath buns and enjoyed hampers from Fortnum and Mason.[3]

Queen Victoria could not keep away, partly because of her pride in her husband's achievements and also because of the fascination of the exhibits. She was presented with numerous artefacts, many of which formed the basis of a fledgling collection that became the Victoria and Albert Museum. Shopping sprees were not dismissed by her

Visitors to the Great Exhibition looking at a cabinet, in lavish high Victorian style, which had been made for the Queen and put on show. *Official Catalogue of the Great Exhibition.* (*Author's Collection*)

either: she bought tea sets from Herbert Minton's company and much more. Unwittingly, in a reverse of the norm, it was the wealthy women who kept the place clean, not their poorer servants. When contemplating the cleaning of the floors, the Exhibition Commissioners realised that the ladies' long crinoline silk dresses would sweep across the floor, brushing dust and debris into the cracks between the floorboards.

The majority of exhibits were from Britain, her colonies and territories; by comparison the presence of the United States at the Great Exhibition was tiny. Other contributions came from Russia, China and Japan as well as the European states. The Exhibition was a great turning point in the development of mass-manufactured goods, marking yet another milestone as the nation's artisan craft economy moved towards one of industrial mass production. Although much of the design of the exhibits was resonant of the high flamboyant Victoriana, this, as Nikolaus Pevsner notes, competed with the simplicity and sophistication of the underpinning technology and engineering of Paxton's futuristic building. Form and function did not flow in empathy with one another.[4]

Thousands of women, of all classes and backgrounds, visited the exhibition where they saw the huge factory machines for cotton spinning, weaving and metalwork. They saw, among other things, smaller machines associated with the new techniques of telegraphy, locks and door handles by Mr Chubb's firm and an envelope-folding machine operated by children for the de la Rue Co. There was the glittering Koh-I-Noor diamond from India, an enormous African stuffed elephant, delicate Paisley patterned textiles from Kashmir, Canadian canoes and furs, Birmingham metalwork, Stoke-on-Trent ceramics, Italian book bindings and British, Austrian and German furniture.

Among these exhibitors were hundreds of women, who were showing publicly that they had ventured beyond the polite arts and were now regarded as designers and inventors alongside men. Their professions were listed as artists, designers, manufacturers and an impressive number of inventors. They included:

- The Dowager Lady Juliana Ashburnham, Hastings Producer. 'A bag of Hops, grown within three miles of the sea in the parish of Guestling, Sussex.'

- Liza Davy, Grosvenor Street, Inventor and Manufacturer. 'New registered riding stays, nursing stays, dress stays, and stays of the usual kind.'

- Mary Galton, Fitzroy Square and Pentonville, Designer and Manufacturer. 'Sofrano standard rose tree, mignonette. Modelled in wax.'

- Maria Gray, Hoxton, Designer and Inventor. 'Group of flowers made of human hair.'

- Mary Mathews, 16 Westbourne Street, Hyde Park Gardens, Inventor. '"Astrorama", with a sketch explaining its use. A concave representation of the heavens, with the apparent diurnal motion of the stars, and real paths of the planets, folding up in the form of an umbrella.'

- Matilda Pullan, from Regents Park, Designer, Inventor and Manufacturer. 'Modern point lace, worked with a common sewing needle.'

- Anne Rayner, 15 Berners Street, Designer. 'Specimens of diamond engraving upon black marble, as applied to tabletops, letter-weights &c.'

- Charlotte Readhouse, Newark-on-Trent, Designer and Producer. 'A Lunar Globe; a model of the moon, giving a general idea of the relative position of the mountains, valleys and plains of our satellite, in relief.'

- Sophia Seltzer, 7 Upper Ranelagh Street, Pimlico, Inventor. 'Chair for spinal curvature.'

- Mary Sykes, Regent Street, Inventor and Manufacturer. 'Corset, weighing only five ounces; the elastic portion made by hand; the mode of fastening and unfastening is new.'

- Elizabeth Wallace, Fitzroy Square, Inventor. 'Slabs of glass, to imitate various kinds of marble, as Sienna, Egyptian Green, red Mona, Bardilla, verda-antique, jaspers, porphyries, serpentines &c; plain and panelled on ceilings &c as pilasters and columns, and inlay . . . When properly backed up, glass obtains all the solidity of stone, and in case of accident is more readily restored by the patent process.' These were vast interior decoration schemes for the new home.

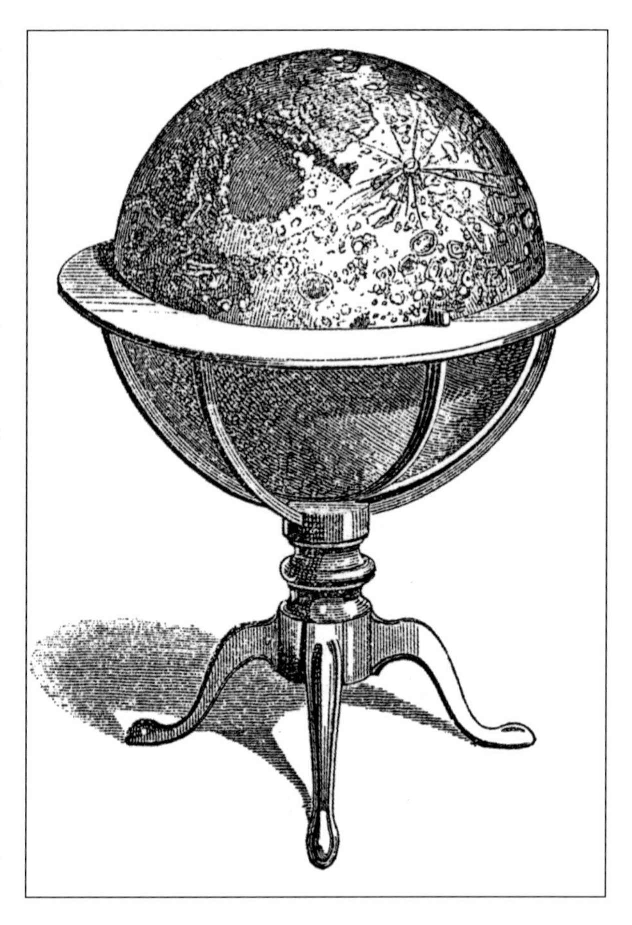

The 'Lunar Globe' exhibited by Charlotte Readhouse. The *Official Catalogue of the Great Exhibition*. (*Author's Collection*)

Madame Roxey Caplin, the corsetière from Berners Street, exhibited her various corsets including the 'Patent improved self-adjusting corsets and child's bodice, ladies' belts &C, constructed in accordance with our present knowledge of anatomy and physiology, and calculated to promote the health and comfort of the wearers.' As a 'Manufacturer, Designer and Inventor' she was awarded The Prize Medal at the Great Exhibition.

It is uncertain if the widowed Elizabeth Dakin had actually invented the Dakin coffee roaster that she patented in 1848.[5] The machine was exhibited by Dakin and Co. of St Paul's Churchyard, in the City of London and the entry states that by cleaning and roasting the coffee in a silver drum its flavour would be much improved.[6]

Due to the success of the exhibition, general acknowledgement grew of the need for inventors' rights to ownership of their intellectual property to be improved. Legislation was passed and a

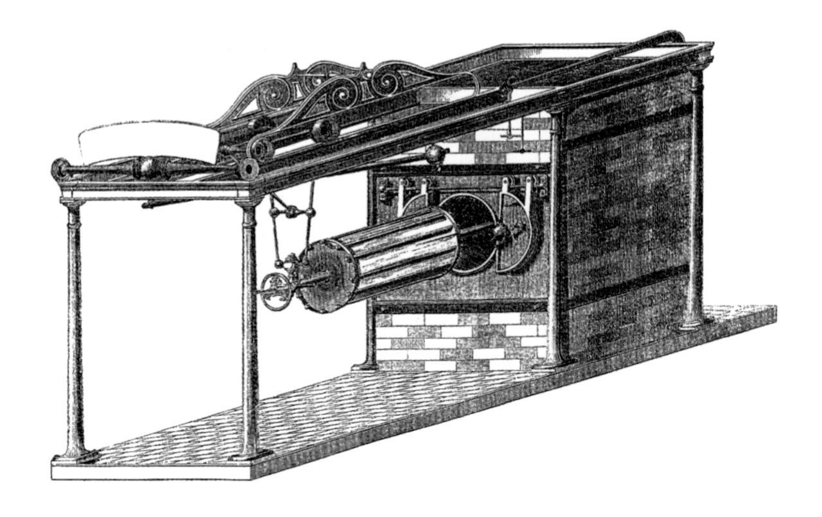

Dakin and Co.'s 'Patent Apparatus for Roasting Coffee in Silver'. *Official Catalogue of the Great Exhibtion. (Author's Collection)*

new Patent Act passed in 1852, much to the annoyance of the free traders who thought that instigators of new ideas should give them away gratuitously to all. No doubt the manufacturers would glean huge profits and the inventors a pittance. This act brought the patenting systems of England and Wales, Scotland and Ireland under one body, the Patent Office, and the cost of registering a patent was lowered. However, the office did not, until 1905, set out to validate authenticity of the idea, preferring to leave that to the courts.[7]

The triumph of the Great Exhibition had impact abroad as well. Suddenly, the need to host an international exhibition became a subject of national pride and more were held in London, Glasgow, Paris, New York, Philadelphia and Chicago: the Great Exhibition had spawned the World's Fair. In Philadelphia a centennial exhibition was held to celebrate 100 years of an independent United States in 1876. But where the Great Exhibition had concentrated on products, in Philadelphia and later in Chicago there was also emphasis on individual achievements. To mark this, women were given their own pavilion with the great-granddaughter of the American scientist and statesman, Benjamin Franklin, as its president. A female engineer operated a large steam engine that powered six looms and a printing press to celebrate women's progress.[8]

It was probably the exhibition in Chicago in 1893 which most closely paralleled for the United States what the Great Exhibition had been to Britain. Originally a tiny fishing village on the banks of Lake Michigan, in the Mid-West, Chicago began to gain commercial importance around 1850, and grew from a population of a few hundred in 1830 to about one million in 1890. Quickly it became the typical American melting pot as Europeans – including Irish Catholics escaping the potato famine and Lithuanian and Polish Jews

The opening ceremony of the Centennial Exhibition, Philadelphia, 1870. (*Illustrated London News*)

fleeing pogroms – arrived to create a polyglot city to rival New York. As Philadelphia had celebrated the centenary of independence, it was decided, 401 years after Christopher Columbus sailed across the Atlantic, to hold a commemorative World Fair in 1893. Like Paxton's building in Hyde Park, the architecture of 400 buildings in beaux-arts and neoclassical styles of the World Columbian Exhibition influenced the future development of architectural style in Chicago.

The energy of the inventors – among them many women – was especially celebrated in Chicago in 1893. In a recent novel, Steven Millhauser describes a fictional scene of an 1890s American city, which could easily have been Chicago: 'on any street corner in America you might see some ordinary looking citizen who was destined to invent a new kind of bottle cap or tin can, start a chain of five-cent stores, sell a faster and better elevator, or open a fabulous new department store with big display windows made possible by an improved process for manufacturing sheets of glass.'[9]

Chicago's enterprising women, many of whom had been pioneers in the drive west, were part of this culture of invention. The furniture store owner, Sarah Goode, became the first Black American woman to patent her invention in 1885, when she designed her cabinet bed. Josephine Cochran invented her dishwasher in 1886. In an effort to acknowledge their abilities the World Columbian Exhibition's commissioners and planners decided this time that the Women's Pavilion had to be designed by a female architect, so a competition was held. Many found this ludicrous, as, like all professions, architecture was male-dominated, with few qualified women practitioners. A further insult was that the winner would not be paid for her work. Out of the twelve qualified applicants the winner was Sophie Hayden, a 21-year-old graduate. Sadly she found the constraints of having to design within the beaux-arts style burdensome and, no doubt receiving much personal attack, she designed only one more building in her career. A golden nail, presented by women from Montana, was to be the last one banged into the pavilion, using a hammer given by the women of Nebraska.

Possibly the planners thought that by honouring women in this way and allowing them such a space they were being progressive. However, their attitudes at times lacked a full understanding and acknowledgement of women's capabilities and what they were achieving. In the British women's section, the constraints as to what they could and could not exhibit were very clearly laid down, and the discrimination was tilted in men's favour. In the handbook for the British section of the exhibition, the suffragist, Millicent Fawcett, chair of the English Women's Work Committee, described the predicament, imbalance and restrictions imposed on the process of selecting a range of work by women, to show the breadth of their achievements:

> The selection of the committee has been strictly confined by the Commission to the work of women; hence nearly the whole field of ordinary industry in which men and women work in co-operation with one another has been excluded. For instance, the textile industries, in which about two-thirds of the persons employed are women and girls, cannot be included because in all these men and women work together. Women's work as designer has also been excluded where, as frequently happens, the design is executed by male artificers. On the other hand, work exclusively done by female fingers, such as the manufacture of handmade carpets, does not find a place . . . if the design has been supplied by a man. In many respects, therefore, the Women's Work Committee have been confronted by difficulties and restrictions . . . The Committee have endeavoured, within the limits just referred to, to carry out the desire expressed to them by Mrs. Potter Palmer . . . of the Board of Lady Managers to gather together at Chicago a record of what women had done.

> The field of Women's Work may be divided roughly under two heads –
>
> 1. Occupations that have been from time immemorial in the hands of women.
>
> 2. Occupations in which women have taken an active part only within comparatively modern times.[10]

Mrs Fawcett continued to describe a situation in which, when women broke new ground and entered untraditional areas, they were able to do their traditional work with more imagination, insight and thoroughness. Due in part to the restrictions placed upon any work in which men had been involved, the range exhibited in Chicago, by women from all over Britain, included a number of displays from the traditional polite arts of lace-making, embroidery, needlework, handicrafts and artistic decorations. There were, however, others from areas new to women. Mrs Fawcett, sister of Elizabeth Garrett Anderson, the first English woman doctor, herself headed the education section that included medical education, and Baroness Burdett-Coutts headed the philanthropic exhibits. In the nursing field there were instruments for surgeons; nursing uniforms; models of procedures and surgical treatments designed by Sister Marion of the Homeopathic Hospital; and 'invalid appliances and furniture designed by nurses'. The Technical College for Women in Liverpool showed photographs of classes, recipes for cooking, sewing manuals, pattern cutting and laundry work. There were also separate sections on the domestic industries of Scotland, Ireland and Wales.

An exhaustive literature section included books of poetry, novels, juvenilia, history and biography, science and miscellaneous, travel and music and exhibited works by Mary Wollstonecraft, Charlotte Brontë, Jane Austen, George Eliot, Millicent Fawcett, Mary Somerville, Eliza Acton, Dr Sophia Jex-Blake and the hundreds of other women who constitute the canon. One spectacular gallery contained hundreds of portraits of

First page of the *Official Catalogue of the British Section*, of the Chicago Exhibition, 1893. (*Author's Collection*)

UNDER REVISION.

ROYAL COMMISSION FOR THE CHICAGO EXHIBITION, 1893.

OFFICIAL CATALOGUE

OF THE

BRITISH SECTION.

FIRST EDITION.

LONDON:
WILLIAM CLOWES & SONS, LIMITED,
13, CHARING CROSS.
1893.

women, from history to the present, as sketches, paintings, miniatures and photographs. In what was a new venture at the Chicago exhibition, a series of debates and seminars on 'women's progress' attracted a lot of interest and provoked huge arguments.

The appearance of one group of women must not be overlooked – that of twenty-three holders of patented inventions. Some of these have already been mentioned but as a group they were:

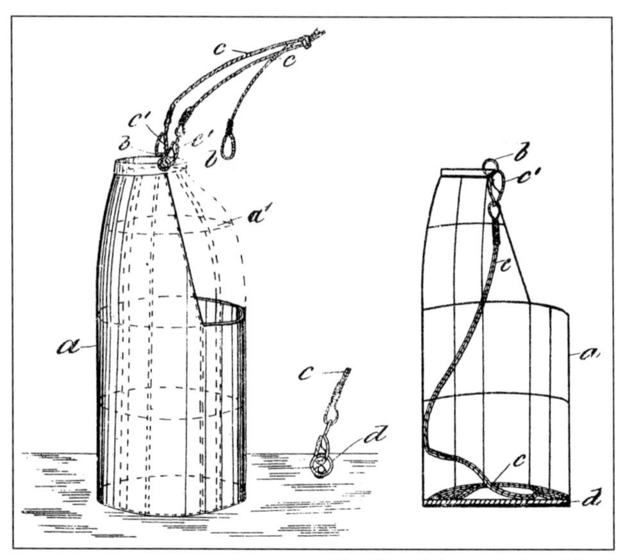

Combined dress stand and fire escape, made of basketwork, by Matilda Barron. Patent.

- Miss M.S. Barron, East Moseley, Surrey. 'Collapsible noiseless coal-scuttle; combined dress stand and fire escape.'
- Mrs E. Barnston-Parnell, Wallington, Surrey. 'Drawings of inventions for extracting gold from base metals.'
- Miss Mary Brown, Clapham, London SW. 'Patent household portable washing copper.'
- Mrs Bewicke Calverley, St James's Park, London W. 'Music folio and stand.'
- Mrs M. Claxton, New Brighton, Cheshire. 'Combined table and bookrack.'
- Mrs E. Dale, Richmond, Surrey. 'Folding mail-cart.'
- Mrs S. Garwood. 'Invention for facilitating the pouring out of bottles.'
- Mrs M. Gladstone, Bury St Edmunds, Suffolk. 'Combined travelling trunk and wardrobe.'
- Mrs E. Grimes, North Walsham, Norfolk. 'Protectors for fingers when sewing.'
- Miss M. Hungerford, Clonakilty, Co. Cork, Ireland. 'Boots and gaiters, shewing new method of lacing same.'
- Miss M. Impey-Lovibond, Ardleigh, Essex. 'Combined sealing-wax holder and seal.'
- Mrs S. Jones, Port Pinorwie, Bangor, Wales. 'Hygienic egg boiler.'
- Mrs M. Kesteven, Hampton Wick, Middlesex. 'Improved Carving Fork.'

Mary Claxton's combined table and book rack. Patent.

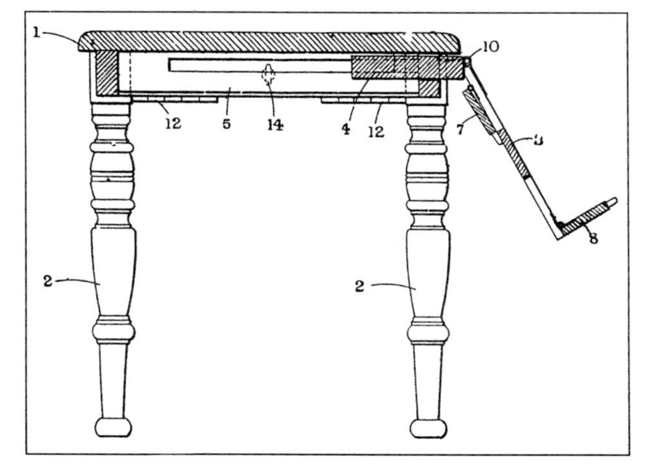

- Mrs S. Mackie, Chancery Lane, London WC. 'Clothes washer.'
- Mrs C. Malcolmson, East Barnet, Hertfordshire. 'Expansible umbrella holder.'
- Mrs McCleverty, Newnham-on-Severn, Gloucestershire. 'Combination skirt stand and table.'
- Miss I. Peckover, Bloomsbury, London WC. 'Sanitary sink basket.'
- Mrs M. Phillipps, Kilburn, London NW. 'Improved ear trumpet.'
- Miss R. Seaton, Sunbury-on-Thames, Middlesex. 'Cleats for holding ends of threads.'
- Miss M. Stephenson, Park Lane, London W. 'Knee music stand.'
- Mrs S. Symons, Guernsey, I.M. 'Adjustable millinery stand.'
- Mrs.F. Tenison, Uxbridge Road, London W. 'Appliance for lifting hot plates.'
- Miss D. Turck, St George's Square, Primrose Hill, London. 'Specimen of painting on textiles.'

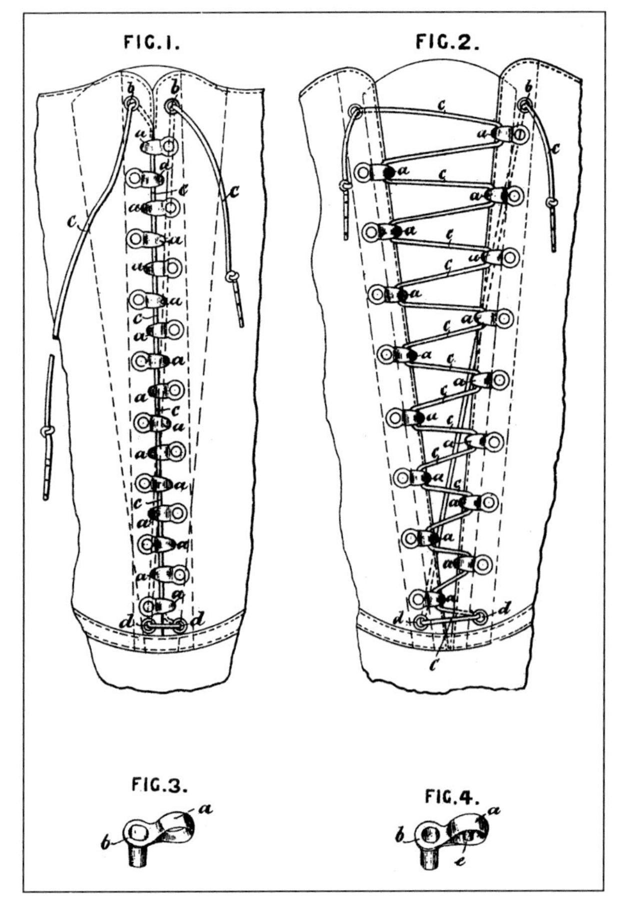

Mary Hungerford's boots and gaiters, shewing New Method of Lacing the Same. Patent.

It is clear that, while the majority of these exhibitors conform to the first of the two categories, it is only Mrs Barnston-Parnell who falls indisputably into the second, that of an occupation 'in which women have taken an active part within comparatively modern times'. That only one woman should be in this category, despite the debates on 'women's progress', is unfortunate. Two, Miss Brown and Mrs Mackie, had taken traditional women's work (the laundry) and combined it with the new technologies, to do as Mrs Fawcett had described, and improved upon them. Considering what else they had achieved and were achieving at this stage, it is apparent that the commissioners' limitations prevented a number of women's inventions from being included. It is an indicator of how difficult the hurdles were for women endeavouring to be accepted in the new technological and design disciplines.

Susan Mackie held two patents for her 'Boiling apparatus for Washing and Laundry Purposes'.[11] With the widespread understanding, by the 1880s, of the need for clean clothes to stop the spread of disease, the new middle-class villa would have been

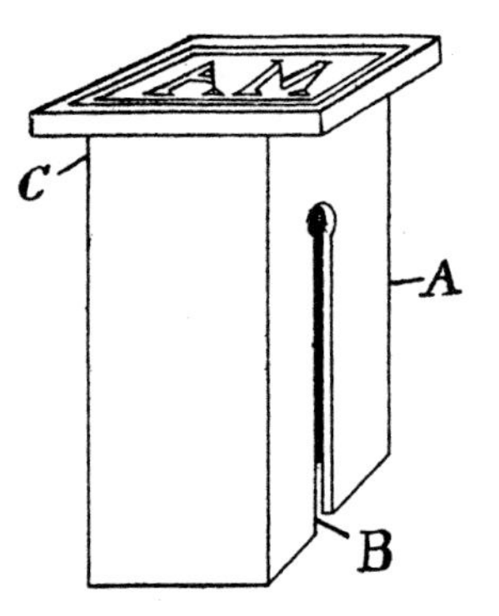

Combined sealing wax holder and seal, by Mabel Impey-Lovibond. Patent.

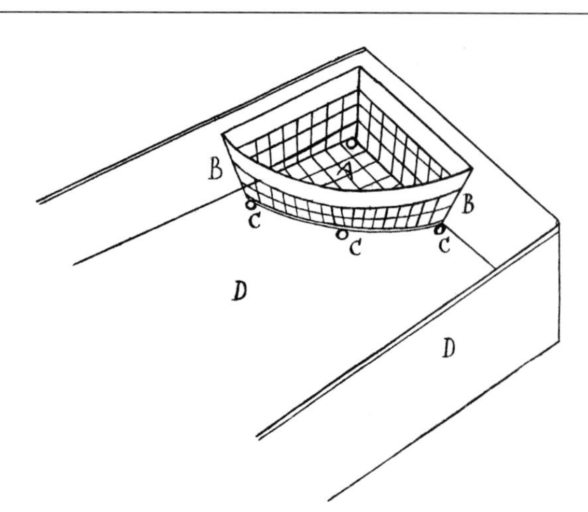

Sanitary sink basket, Isabella Peckover. Patent.

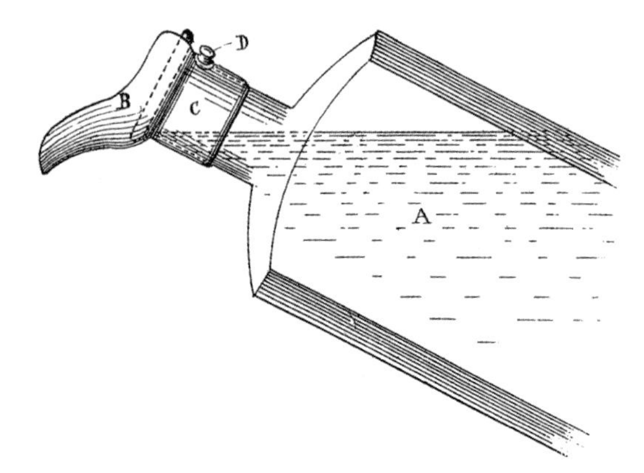

Invention for facilitating pouring out of Bottles, Sarah Garwood. Patent.

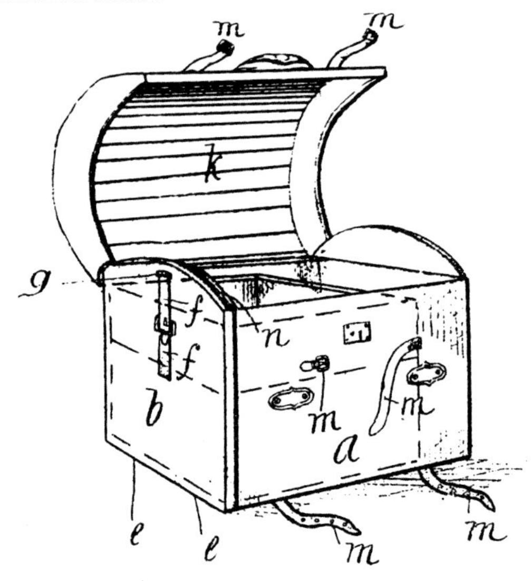

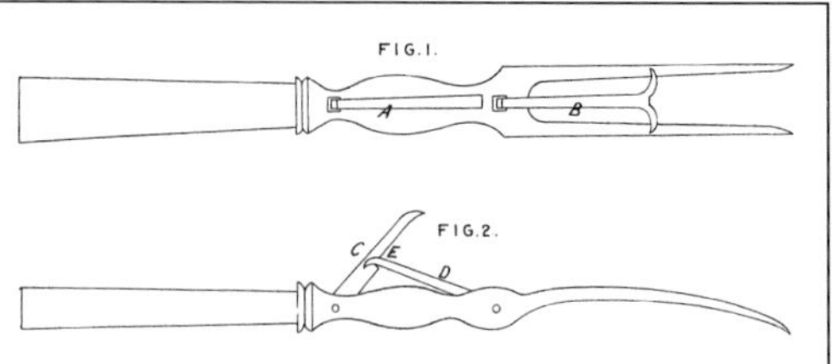

Improved carving fork, Mary Kesteven. Patent.

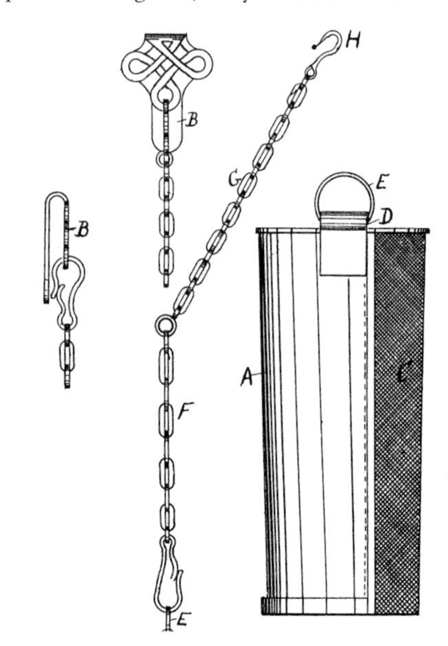

Expansible umbrella holder, Catherine Malcolmson. Patent.

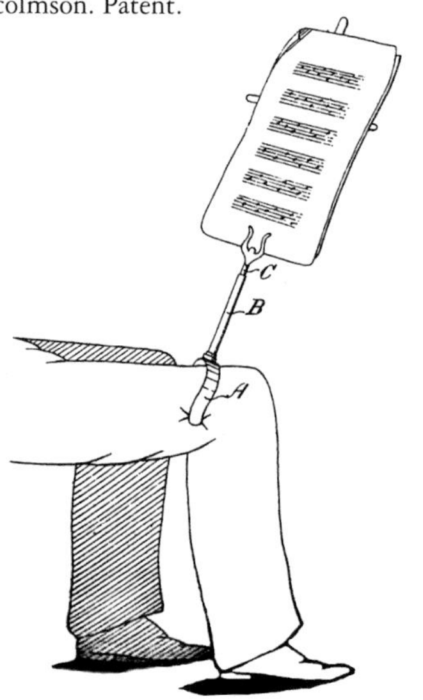

Left: Combined travelling trunk and wardrobe, Mary Gladstone. Patent. *Above*: Knee music stand, Marianne Stephenson. Patent.

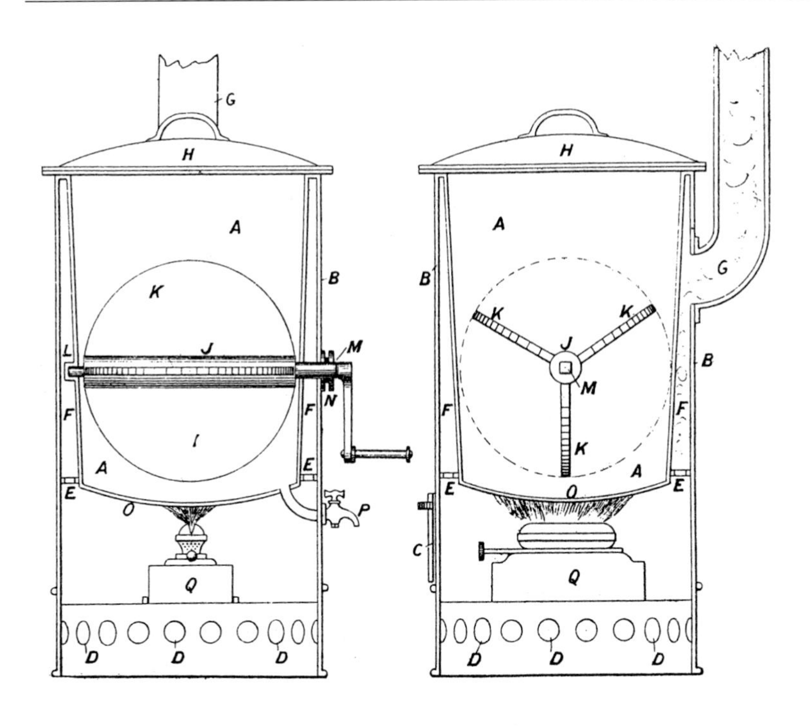

The Patent Household Portable Washing Copper invented by Mary Brown. Patent.

Improved ear trumpet, Marian Phillipps. Patent.

Below: Susan Mackie's Clothes Washer. Patent.

designed with space for a laundry, operated by housemaids. However, such luxury was not for the poor. Instead, the new public wash houses and baths not only incorporated bathing facilities but also laundries, operated by women, to which others brought their dirty washing. Susan Mackie states that her washing machine could be used in a domestic or public laundry. This was a double hemispherical boiler with a space between the inner and outer walls through which the hot water, pressurised by the steam, would flow. At the top it would gush out in jets through holes into the inner boiler.

Elizabeth Barnston-Parnell described herself as both an inventor and a metallurgist. She applied for numerous patents; not all were accepted, but three did proceed, two in 1889 and one in 1891, the drawings of which were exhibited at Chicago. She was born in

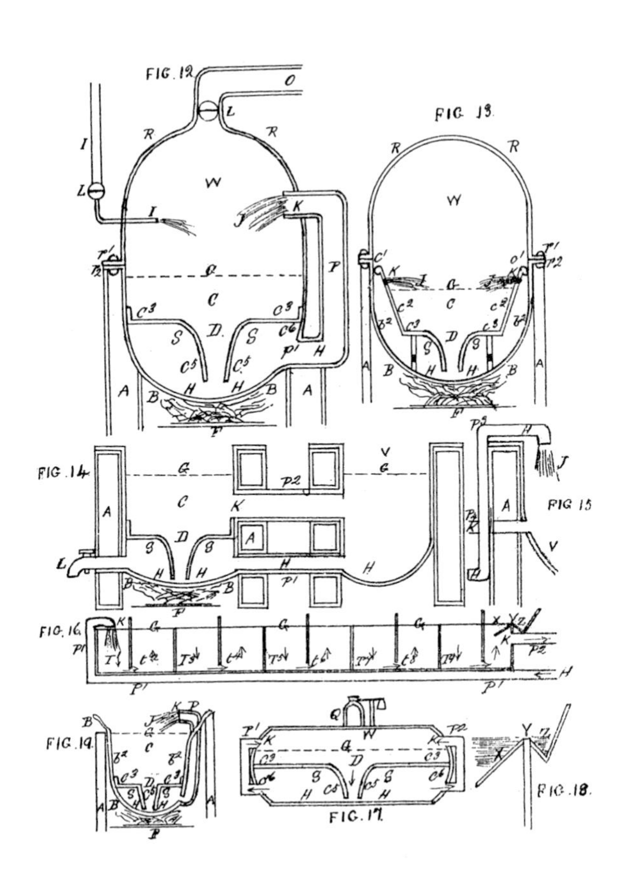

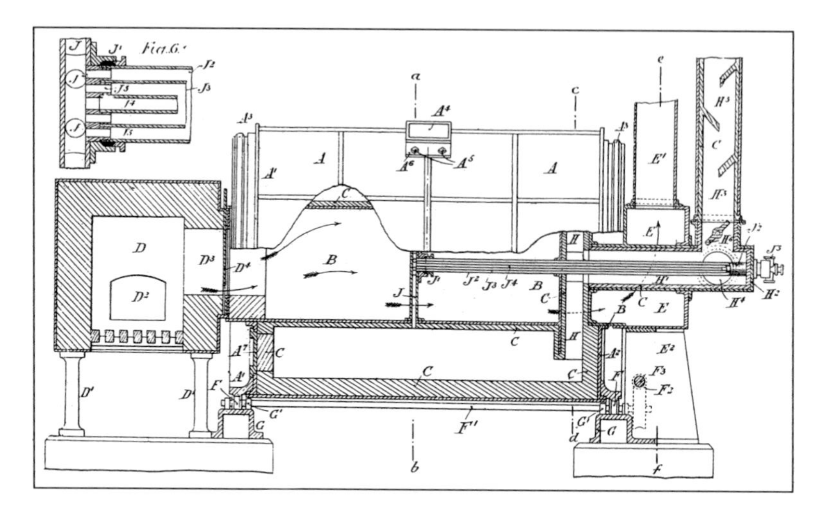

Drawings of inventions for Extracting Gold from Base Metals, devised by Elizabeth Barnston-Parnell. Patent.

Carmarthen in 1837 and later lived in Sydney, Australia where two daughters were born. However, in 1889, the date of the first patents, she was based in Kent.

The aim, in her 'Improvements in the Treatment of Ores for the Recovery of Metals and in Apparatus thereof'[12] was to reduce the cost of extracting gold and other metals from ore with appropriate, cheaper machinery rather than the stamps which were in current usage. The description of what Mrs Barnston-Parnell called a 'mechanical' process conjures up images of heat, sweat, dust and personal danger in a late nineteenth-century factory:

> I take the quartz or other stones or ore and in a suitable furnace I heat it to the necessary degree usually a fair red heat; I then place the heated material in water or apply water to it with the result that immediate disintegration takes place and the hard blocks of say quartz are at once reduced to a soft consistency so that they could be crushed with the heel of a person's boot . . . I employ any suitable form of furnace for heating the air which is blown or sucked by any suitable means such as a fan or blower through the hollow shaft and webs of a worm which revolves inside any suitable form of casing preferably of such length that when the ore which is drawn along by means of the worm has reached the end of the casing remote from that at which it entered it shall be sufficiently heated for my purpose.

The hot material then fell into a tank of cold water and then was transported 'preferably automatically as by a bucket or other conveyor' to a drum or brick chamber where it was dried using hot air which had escaped from the first machine. Its immediate drying meant that it could be promptly passed, again on an automatic conveyor, to a crushing machine. Here Elizabeth Barnston-Parnell's preferred rolls would be used, instead of the commonly employed stamps, because the substance was now softer than that resulting from previous practices.

Mrs Barnston-Parnell was also aware of the benefits of her process in places like Canada, Russia and Siberia where the cold and freezing winters would stop the extraction of ore. Here, by using her process, they could melt ice and snow and use it for cooling purposes. In conjunction with this process were the 'Improvements in or relating to Calcining Furnaces'[13] in which the furnace, made of cast iron, had six chambers arranged around the tube conveying the heat from the fire grate. Air, steam or gas passed through pipes into the chambers. The whole furnace could be rotated with a worm gear.

No doubt the Commissioners for the World Columbian Exhibition thought their efforts to include and accommodate women were laudable. But, despite the designated pavilion, the framework for selection they imposed was out of step with women's newfound confidence, capabilities and aspirations. All the political and technological revolutions of the nineteenth century had resulted in changes to the relationships between women and men. The wider world had become accessible with the new transport and communication changes, and inevitably broadened women's horizons. Although they were prohibited from exhibiting all that they would have wished, fierce debates concerning all the current women's issues did take place in that pavilion at Chicago. Appropriately, they were all reported back and published in the *Illustrated London News* by its correspondent Mrs Fenwick Miller – with advertisements for the latest model of the new Juno Ladies Bicycle appearing on the facing pages.

CHAPTER TEN

Running the Company

Despite all the obstacles in their paths, women succeeded not only as inventors and designers but as entrepreneurial business people too. Using their inventions and ingenuity, good ideas were built upon to create successful companies; beneficial contributions were made to deceased husbands' firms and new businesses started.

Mouldings and friezes, the products of Mrs Coade's eighteenth-century Lambeth factory still adorn the façades of many grand buildings around Britain. Mme Roxey Caplin's aim was to promote a healthy lifestyle, which she did by running her gymnasium, designing award-winning corsets and writing books on good posture and the workings of the body from her premises in Berners Street. The philanthropic and indefatigable Amelia Lewis designed, manufactured and marketed her stoves as well as editing and publishing *Woman's Opinion* and the booklets to advocate her belief in healthy living alongside the detrimental effects of alcohol. In Germany, Käthe Kruse's dolls, Margarete Steiff's soft toy animals and Melitta Bentz's coffee filters became the bases for hugely successful, global companies, all of which have retained these women's names. Beatrix Potter's early understanding of character merchandising continues to enrich the profits of her publisher, Frederick Warne, and the enjoyment of children. Many women, like the American Martha Coston with her pyrotechnic signals, improved upon their deceased husband's designs.

Having inherited the company, the need for close involvement could be crucial to its survival. For years anchovy paste was made by the Burgess family in their shop at 107 The Strand, by the Savoy steps, in central London. When they first settled there, in the 1760s, Thomas Coutts had just bought the premises across the road for his new bank. It was here that for three generations the Burgess family ran their Italian warehouse: a nineteenth-century equivalent of a modern delicatessen. In it they sold not only their own anchovy paste in ceramic pots but also other delicacies, including smoked salmon, Welsh oysters, herrings, Dutch beef, reindeer tongues and hams from

Westphalia, Portugal and Westmoreland. William Burgess, son of John the founder, was widowed with six children when he married his second wife, Elizabeth, twenty-six years his junior. Together they had one son, Arthur. When William died, aged seventy-two, at Dover in 1852, the firm passed to Elizabeth and his sons, her stepsons, Arthur not yet being of age.

Although the Burgess food emporium was well established, it was the anchovy essence, first made by John Burgess in the 1770s, which was the company's main claim to fame. In an age before refrigeration and vacuum canning, the unique Burgess preserving methods, which were developed and refined over the years, meant that their anchovy essence lasted for a long time. As a consequence, it was very popular with travellers abroad and consignments were sold to Nelson's naval fleet, Lord Byron and others venturing overseas. Years later Burgess's Anchovy Essence would be an important comestible in Captain Scott's rations on his Antarctic expedition. Royal recognition was confirmed in 1852 when Queen Victoria awarded the company a warrant.

The Burgess Anchovy Essence Trademark. *Trademarks Journal*, 1876.

When the widowed Elizabeth Burgess took the reins of the company she was faced with two major problems: firstly that of competition from imitation preserved anchovies and secondly, differences of opinion with her stepsons about the management of the business. In an attempt to stifle any bogus products, and having taken legal advice, she decided to patent her method to preserve anchovies. This was aimed to retain the fish's full flavour and its delicate meat would be preserved for an 'unlimited period'. In Mrs Burgess's method

> the anchovy is to be scraped or rubbed with a blunt instrument, so as not to injure the skin of the fish, and yet to divest it of all salt and other extraneous matters, and the heads and tails taken off; the fish is then washed in cold spring water, and afterwards carefully wiped with linen cloths, and then placed in bottles, jars, or other suitable air-tight vessels, with spiced malt vinegar or best Lucca oil with the addition of bay leaves, to add to the flavour. The effect of this is that the flavour and all the properties of this peculiarly delicate fish are retained for any length of time.[1]

At the same time, Elizabeth Burgess oversaw the running of the company and was greatly concerned by the precarious state of its finances, the debts left by her husband, difficulties with cash flow, the behaviour of one of the stepsons and the coming of age of her own son. She had long and detailed meetings with her solicitor from the City firm of J.H.N. Linklater and Hackwood, and had, at times, to make some hard financial decisions. She steered the company through difficult times and, importantly, protected one of its most important assets, the method of preserving anchovies; yet, she is barely mentioned in any of the company histories even though it remained within the Burgess family until the 1930s.[2]

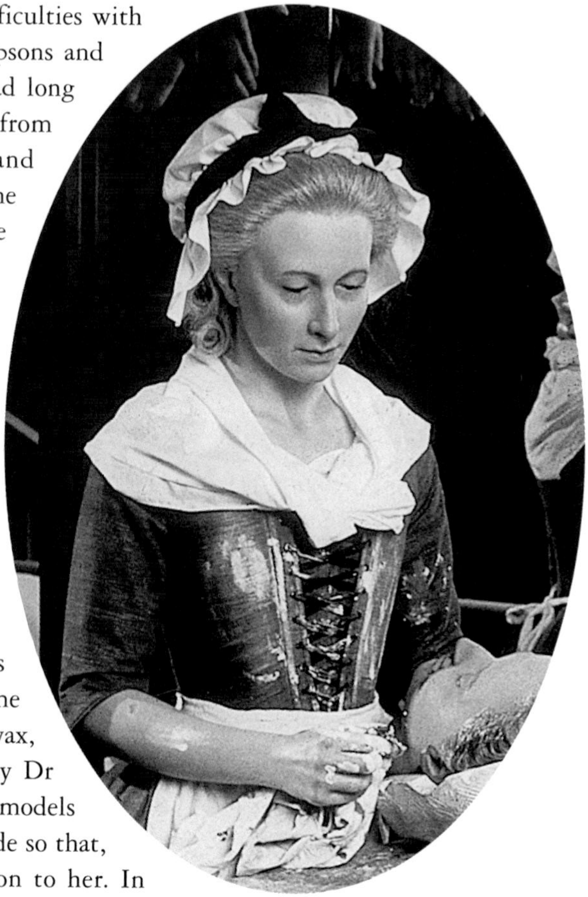

Marie Grosholtz's wax models would add a new dimension to mass entertainment for England's middle classes during the early nineteenth century. She was born in Strasbourg in 1761 and moved to Paris at the age of five, when her mother became housekeeper to a doctor and modeller in wax, Philippe Curtius. Marie was fascinated by Dr Curtius's method of making life-like, wax models of famous people and quickly learnt his trade so that, on his death, he bequeathed the collection to her. In 1795, Marie married an engineer, François Tussaud, but quickly this became a desperately unhappy marriage. As a means of escape, she fled, with her two sons and the collection of wax models, to London in 1802.[3] Waxwork replicas had been popular attractions at fairgrounds but it was Mme Tussaud who, by bringing her characters to the new assembly rooms in places like Brighton and Bath, made them an attraction for the wealthy middle classes where they would come to stare at her latest creations while her sons played music in the background.

Madame Tussaud at work. (*Madame Tussaud's Archives*)

When Mme Tussaud was still in France she had made tableaux, in wax, of scenes of the French Revolution; these included Louis XVI and Marie Antoinette eating a dinner and the Queen getting ready for bed. But it was Queen Victoria's accession to the throne, in 1837, that gave her the opportunity to pursue her beliefs in the monarchy and family values, which would have found a welcome audience among the aspirational middle and working classes. Having toured the collection around the country for thirty years she eventually settled it, and herself, at a permanent base in Baker Street just around the corner from the site the company moved to and continues to occupy in Marylebone Road. Even before photography and mass-market journals became widely available, Mme Tussaud, by understanding the fascination the public has for the rich and famous, gave them the opportunity to believe that they really were mixing with them. This myth still pervades when, on a visit to Madame Tussaud's, people play with fantasy and reality, having their photographs taken alongside the model of a famous person to maintain the illusion they have been in the presence of a celebrity. When Marie Tussaud died in 1850 at the age of eighty-nine, not only had she created a wonderful popular entertainment but she also laid the foundations of a flourishing company that continues to enthral and excite with its attractions all over the world. But it is still Madame Tussaud's in London, which, with its intimate wax models, is its main attraction.[4]

Even the introduction of the Married Woman's Property Act in 1884, the future was still fraught with problems. Old prejudices and traditions are hard to discard and, for women, entry into the male domains of business and industry was no exception. The pottery industries of Stoke-on-Trent provide an example. In the eighteenth century Josiah Wedgwood, and in the nineteenth, Herbert Minton, had regarded the designer's role as fundamental to the success of their products. Employers in the Potteries encouraged women to attend courses at the local art school when it opened in the 1850s, but having learnt their specific skills of ceramic painting, they were only allowed to decorate the pots; all design work and aspects of technology for production was prohibited to them. Clarice Cliff, Charlotte Rhead and Susie Cooper, employed as designers in the Potteries, only appeared in the twentieth century.

From the 1880s there was a strong revival of the craftsman artisan, relating to the period before the Great Exhibition and its affirmation of the new industrial age, its champions including John Ruskin, William Morris and the Arts and Crafts movement. Although rightly believing that traditional crafts would be threatened with the onset of industrialisation, they were unable to countenance the possibility that anything pleasing might emerge from an industrially designed, mass-produced artefact. While the majority of their one-off pieces were beautifully crafted, they were very expensive, precluding their purchase by those from the poorer classes. It was usually highly skilled women who laboriously made them, using traditional techniques and working to designs dictated by men. It is no wonder that many women educated at art and design schools, like the potter Dora Lunn, were faced with no alternative but to establish their own studios. These were, inevitably, small, craft-based businesses producing short batches of well-designed pieces. Speculation can only suggest that many of these qualified women would have preferred to work in the exciting, design studios of industry rather than in small studio workshops.

Earthenware dish designed and made by Dora Lunn in her studio. (*Victoria and Albert Museum*)

Dora Lunn was born in Sheffield in 1882 and came to London as a child. She took classes in ceramics at Camberwell School of Art in South London, and at the age of twenty-six was awarded a certificate in ornament and design by the Royal College of Art. From here she went to work in her father's silversmith and ceramics studio but decided to exert her independence and set up her own pottery studio. Almost single-handedly she transformed a rundown tailor's shop in Ravenscourt Park, West London. In a hive of activity, wooden boxes were made into wall cupboards to store glazes, tables into workbenches and shelves put up for the finished pieces. Copious notes remain which describe Dora Lunn's love and respect for her work, the determination to set up a pottery and the other women she employed. Her vision to produce small, simply designed pots was no doubt aesthetically influenced by the Arts and Crafts movement, but was also governed by the fact that she could only afford a tiny kiln. Any financial profit she received was minimal until an anonymous patron appeared, making it possible for her to make bigger pots in a new, larger kiln. What is outstanding about Dora Lunn is her abundant self-confidence and belief in her designs and industry; she never had any doubt that she might not succeed. By 1917 she was the first woman potter to exhibit at a British Industries Fair; lectures followed and for years a vast range of simple yet beautiful bowls, plates and jugs were produced, in the studio in Ravenscourt Park, for the domestic market.[5]

A couple who, by combining their skills, produced richly coloured ornamental metalwork was Edith and Nelson Dawson. Edith had developed her awareness of colour as a watercolour painter before their marriage in 1893. This intrinsic knowledge and appreciation of colour enabled her to produce an array of innovative, decorative enamelwork to adorn his silverware. In separate studios at their Chelsea home, his beneath hers, they each developed new techniques and designs. Edith used new methods like *cloisonné*, in which a wire outline was soldered on to a base to make a raised line into which enamel colour was poured. Together, Edith and Nelson Dawson produced a great variety of work, including candlesticks and tableware for the domestic market and screens for embassies, law courts and churches.[6] According to Anthea Callen, although Edith received little public recognition for her work, by 1897 she was in charge of all the enamelwork design and a small labour force to meet the rapid demand for their work. In 1906 she published *A Little Book on Art – Enamels* in her own name. Nelson was a member of the Art Workers Guild, a benefit denied to women, although both belonged to the Arts and Crafts Exhibition Society. In 1900 they exhibited 125 pieces at the London Fine Art Society Fair. But enamelling is very dirty work, giving off terrible fumes, which, combined with overwork, caused Edith's health to suffer so that she had to retire from it in 1914. She died in 1928 at the age of sixty-six.

By the 1880s, the nature of retail marketing changed partly as a result of the promotional aspects of the international exhibitions and the abundance of mass-produced goods. Alongside small retail shops the new phenomenon of the department store appeared. Whiteleys had opened in Bayswater in 1863 and the American Harry Selfridge had opened his enormous store in Oxford Street in 1909. In these scrumptious buildings, almost miniature versions of the international exhibitions, shopping became a major activity of middle-class women. Traditionally the corsetière, milliner and dressmaker had operated from their own workshops to produce one-off, bespoke items. Now, with mass production, even the corset was factory-made. In Luton the straw hat workshops had become small factories, many of which were run and owned by women who had become quite a vociferous force; they were impossible to unionise, insisting that their working hours had to be flexible around family and other commitments. Even though aspects of straw plait work were unpleasant they did take terrific pride in their work and adapted their skills to design and make felt hats for sale in the fashionable department stores.

Small businesses like cafés and kiosks appeared in the streets around the department stores to meet the needs of these unchaperoned women, new to the commercialised city.[7] Shopkeepers too became more sophisticated at displaying and advertising their wares. Hedwig Godefroid[8] from Berlin introduced her electric till which was both a cash register and receipt machine, essential as more people, other than the owner, were now working in the shops and responsible for the money. Elizabeth and Charles Drake,[9] both confectioners, designed a simple and clean-lined display stand for cakes and sweets for a shop window. This consisted of a hollow, enamelled tin pedestal upon which a convex-shaped display tray stood at an angle, the inclination making its contents easier to see.

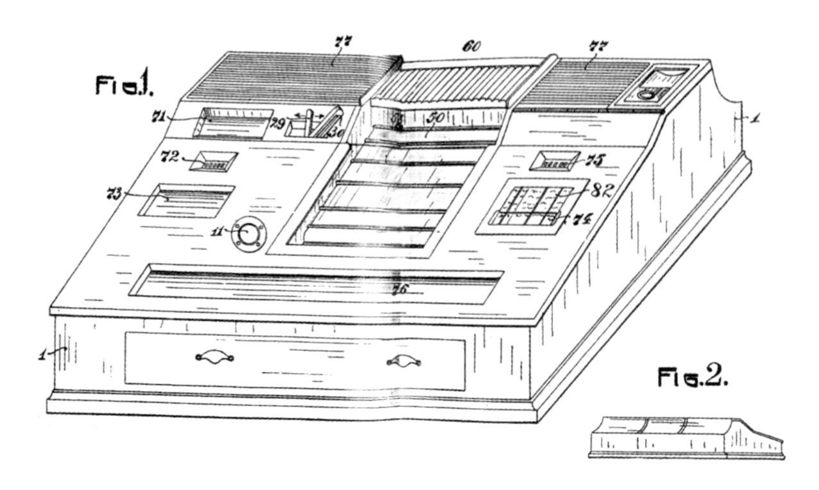

The front of the till designed by Hedwig Godefroid. Patent.

Advertising a business is essential to its success and was understood by Julia Richards and her daughter Mary. They had their own company, bottling mineral water in Newport on the Isle of Wight. The 57-year-old Julia patented her almost subliminal method to advertise or display messages in soda water syphons.[10] The inner glass tube, which could be clear or opaque, was either embossed or engraved with an advertisement or message. If opaque glass was used it could be white or coloured which would, 'besides offering opportunity for the display of devices, advertisements, and other announcements, also enhance the brilliancy of appearance of the contents of the syphon'. Florence Slater, Trading as C.J. Cuthbertson in Cheapside, City of London, produced an 'Automatic Apparatus for Causing the Extension and Contraction of Opera Hats and the like for Advertising Purposes'.[11]

Years before in Boston, Massachusetts an inveterate inventor, Margaret Knight, had devised a machine to make flat-bottomed paper bags. While working at a factory in Maine she had found a method to shut down machines which did not work properly and so prevented many accidents to the workers. It was while she was working for the Columbia Paper Bag Co. in Massachusetts during the 1870s that she invented and patented her machine for flat-bottomed paper bags.[12] Problems arose while she was refining the design and she discovered her idea was being plagiarised and patented by someone else. A furious Margaret Knight fought Mr Annan, her adversary, who claimed, in his defence, that a woman could not possibly have designed it. Eventually Margaret won, patented her invention and successfully founded the Eastern Paper Bag Co. Her machine used a 'continuous tube of paper fed from a roll over a former, and cut, folded, pasted and delivered'. It was then

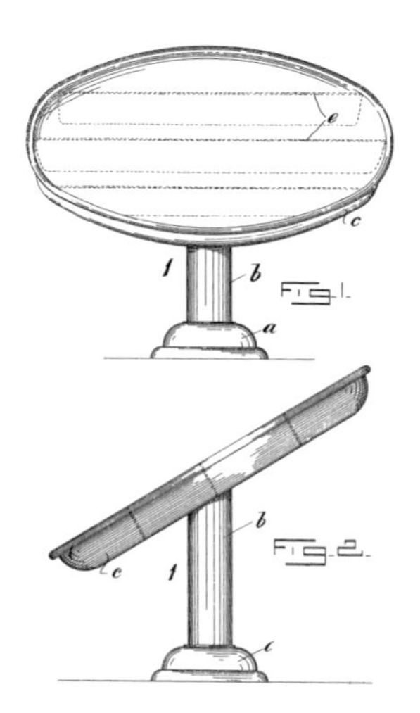

Elizabeth and Charles Drake's modernistic shop display stand. Patent.

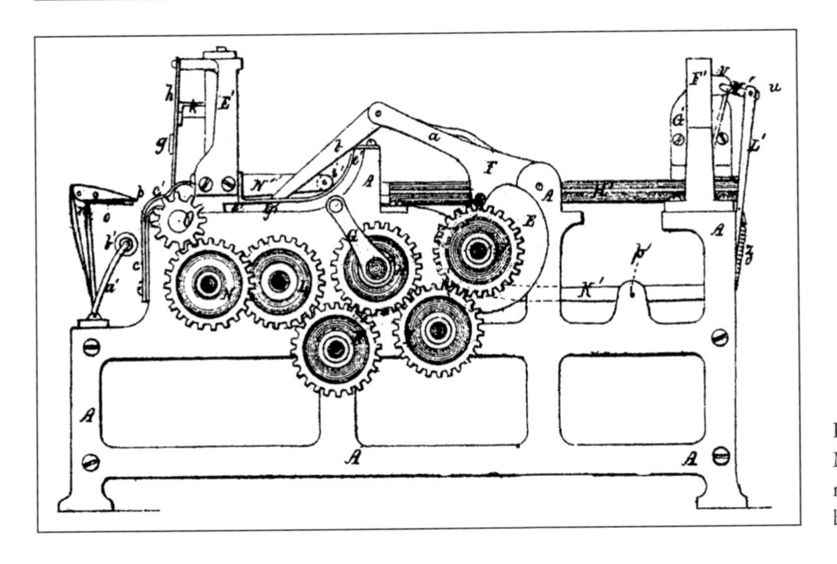

Part of the diagram for Margaret Knight's machine to make flat-bottomed paper bags. Patent.

grooved, folded and cut to form a rectangular, flat-bottomed paper bag; and these same principles are still in use today.[13]

By the early 1900s there were also changes in how women referred to themselves, some having trading names, others professions, giving an indication of their involvement with a company. In Scotland the Spiers family, comprising Jane and Maria and Mary Jackson, née Spiers, were Steel Plane Manufacturers; along with their foreman, William McNaught, they designed a woodworking plane.[14] The base and sides were made of steel with the blade fixed into a bed of malleable iron, steel or wood. Its handle was made of rosewood and a knob on the front meant that the whole plane could be pushed with the right hand and steadied with the left.

Some women became theatre manageresses and lessees; others managed laundries, telegraph exchanges and offices. In all these areas were women who invented and designed numerous devices to help them in their work, including exit signs for the emergency evacuation of a theatre, seating systems for the audience and safety devices for the laundry. Mary Taylor, a laundry manageress in Huddersfield, was one. Public health concerns had made large public and private laundries common in most towns to enable working women and housekeepers to bring their dirty linen to be washed, dried and ironed. Mary Taylor designed a protective guard for hands while using an ironing

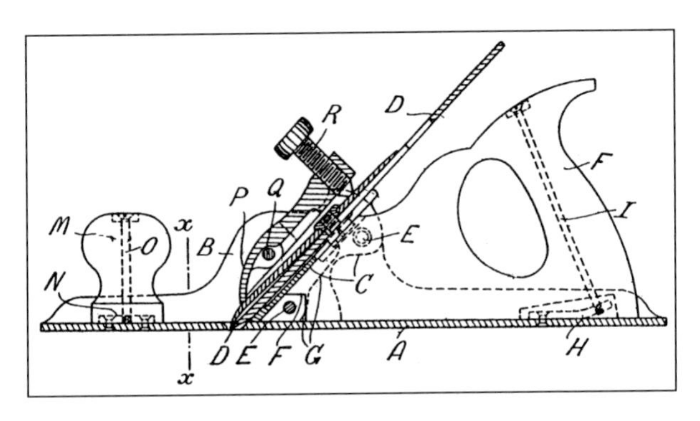

The Spiers family's woodworking plane. Patent.

machine.[15] It was specifically designed for machines to iron collars in which a lethal rotating ironing roller was used with a table:

> In employing this class of machine, the ironing is attended with considerable danger, as unless great care is exercised in feeding the articles to and manipulating them on the reciprocating table, the hands of the attendant are carried forward beneath the ironing roller . . . and thus serious injury may result therefrom.

The guard was a horizontal bar, mounted on to adjustable vertical bolts immediately in front of and parallel to the ironing roller.

Despite these developments, the world of bigger business remained elusive and stubborn to women. They could work on the floors of factories and occasionally become managers, but the running of large companies remained the domain of men, even though women had proved the commercial viability of many of their inventions. These, combined with the complex management and problem-solving skills required for the successful running of the home, were overlooked.

CHAPTER ELEVEN

Self-Confident Expressionism

The lure of the city grew ever greater. Women, now completely free of chaperones, and travelling on bicycles, buses and in London the new underground system, could not keep away.[1] The prospect of work, the new shops and department stores, the theatre, music hall and other entertainments enticed them to it. They fantasised and reinvented themselves in department stores such as Whiteleys, Selfridges and the Liberty emporium on Regent Street, offering its unique mixture of exotic goods from India and the east. Judith Walkowitz describes these women, fascinated by the excitement and frenetic activity of the city, which until then was unknown to them:

> In the controlled fantasy world of the department store, women safely reimagined themselves as flâneurs, observing without being observed, constructing dreams without being obliged to buy . . . By the 1880s, a set of ancillary services outside the department stores – inexpensive tea shops, eating establishments, public lavatories, and cheap public transportation . . . serviced female consumers, allowing them to enter the city centre and to enjoy the new public leisure of shopping.[2]

After her long period of mourning for Prince Albert and the death of Queen Victoria in 1901, the accession of her roguish eldest son, Edward VII, brought an air of liberation which permeated British society, and enjoyment and fun became more openly permissible. More than ever before women were in sole charge of their lives, although the vote still eluded them. For these women, admission to the city and independent travel brought both excitement and danger that demanded that they took control of their personal safety. Bertha Ortell had devised her 'Combined Pocket and Garment Supporter' in which to hide valuables and Louisa Llewellin had knitted her 'Gloves for Self Defence' with which to scratch and scar the face of an assailant. In 1910 Bertha Stahlecker from Wurtemberg in Germany devised a door-fastening signal, which if pushed or leant against would emit a mechanical or electric signal to operate an alarm,

making it ideal for single women living in lodgings.[3] Another German woman, Clara Kunze from Breslau, designed a hook that locked the coat or jacket on to it, thus securing it from thieves.[4]

While the theatre and opera had been cultural venues of the growing middle classes and upper classes for decades, from the 1860s the music halls had become a mainstay of popular entertainment and in 1875 there were more than 300 in London alone. Here women and men would watch raunchy acts, full of sexual innuendo, performed by Marie Lloyd, Lillie Langtry and others depicted in the paintings of Walter Sickert. Women became the licensees of theatres. Together, Amelia Horne, who was married to an engineer, and Kate Santley, a single woman who was also the lessee of the Royalty Theatre in Soho, designed safety signs to evacuate a theatre quickly.[5] Using luminous paint they displayed a series of signs on a continuous band of material that could be hung around the theatre. 'Way out', 'exit' and 'out' each with an indicator arrow and a pointing hand were easily visible in the dark. Caroline Claypole traded as the Armstrong Automatic Manufacturing Co., in Wilton Street in the City of London, and in 1904 she registered her patent for 'Self-closing or Retractible Seats or Chairs',[6] which would have been used in one of the new theatres being built at the time. In line with the need to advertise and promote their businesses, Florence Boggis, a dancer who was 'professionally known as Mlle de Dio', designed a method to display her stage sets and advertisements.[7]

The activities of actresses on stage mirrored women's needs and their desired place in society. To the outrage of some, Sarah Bernhardt had appeared on a Parisian stage in a trouser suit; as elsewhere, in everyday life, women were trying to negotiate riding a bicycle in a divided skirt so as not to reveal their ankles. In these heady days of Edwardian permissiveness they began to smoke in public, a provocative activity that was reflected and no doubt encouraged by photographs of actresses like Lillie Langtry. The era of seduction and cigarette smoking had arrived. Until the mass manufacture of cigarettes in the 1840s, mainly upper-class men had

Marie Lloyd. (*Private Collection/Bridgeman Art Library*)

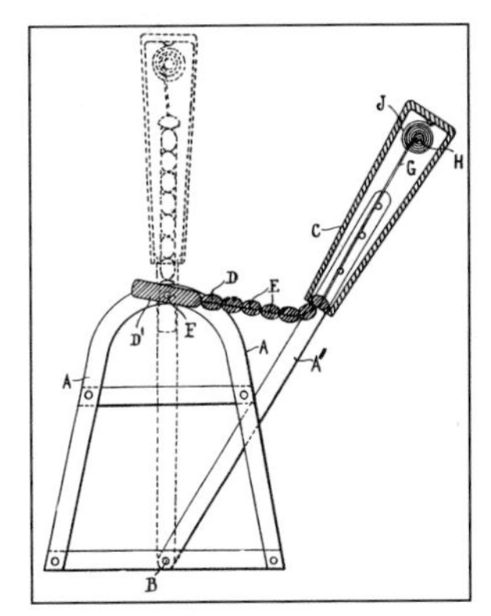

Caroline Claypole's self-closing theatre seat. Patent.

Opposite: Shopping in Southampton. (*Illustrated London News*)

Suffragettes chained to railings outside 10 Downing Street, 17 January 1908. (*Illustrated London News*)

smoked tobacco, in cigars and pipes. In the 1860s Golden Virginia tobacco arrived from the USA and was used by companies like Wills in cigarettes intended to be consumed by men. No Victorian dinner party was complete without the parting of the sexes after the meal – the men to smoke and the women to retire to the drawing room, to be joined by their refreshed escorts later. As the 1880s had produced posters of the freethinking and independent woman riding a bicycle, by 1905 she was just as sophisticated but now smoking a cigarette. Suddenly a woman, cigarette in hand surrounded by curls of smoke, was the height of sophistication and charm. In 1897 Margaret Montgomery Grant, an artist from Aberdeen, developed an interesting pipe for smoking tobacco.[8] To avoid contact between the fingers and the tobacco in the pipe she made a 'cartridge' of thin, waxed muslin to hold the requisite amount. Once smoked the pipe would emit an aroma due to her having dipped one end of the cartridge in 'spirits of wine . . . or similar liquid' before placing it in the bowl. She believed that her method would also mean that 'the evil effects of nicotine are greatly diminished, the ash does not fall about and a cool and refreshing smoke is provided'. Amy Booker, a 39-year-old married woman from Kings Langley in Hertfordshire, designed a pipe made of gold, silver, copper or brass.[9] The widespread practice of smoking required ashtrays, and Henrietta Pearce from Berners Street designed one to attach to the side of a table – which she found especially useful when playing billiards![10]

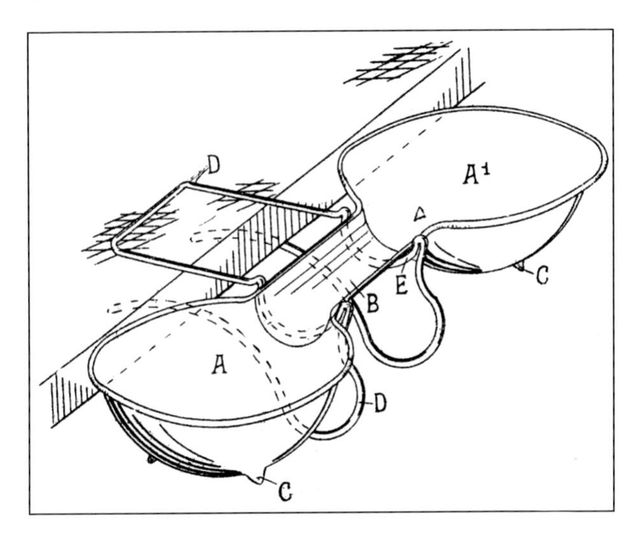

Pearce's ashtray to clip on to the side of a snooker table. Patent.

Corsets, too, had become more relaxed for bicycle riding and a generally freer lifestyle, as the tight-laced, all-in-one garment gave way to the bra and suspender belt. The approach to the all-in-one corset, too, reflected the raunchiness of the late Victorian and Edwardian era. Even the style of drawing in the diagrams emits frivolity and fun absent in earlier illustrations and the flat diagrammatical representations were replaced by more relaxed drawings of corsets and undergarments on female bodies, as in that by Maria Pressaud from Eaton Square in London.[11]

The automatic musical instrument had reached many homes, providing easy renditions of popular tunes, like the one designed by Florence Heppe, a manufacturer in Philadelphia.[12] A mechanically turned web of perforated paper hit the correct marks on a tracker bar, causing the hammers in the instrument to be struck and the appropriate sounds emitted. The outer casing looked like a real piano or organ.

The Bank Holiday Act of 1871, the expansion of the railways and Thomas Cook's packages all contributed to the opportunity for holidays, access to travel and the growth of the seaside resort. Brighton had long been a royal favourite. Even though Queen Victoria loathed it – and sold off the Pavilion to pay for the building of Osborne in the 1840s – it continued to attract visitors from London. Other resorts like Blackpool and Southport evolved for those from Lancashire, and Scarborough served Yorkshire. With their beaches, promenades and fairgrounds they provided hours of carefree fun for late Victorian and Edwardian families. In 1909, a widow, Jane

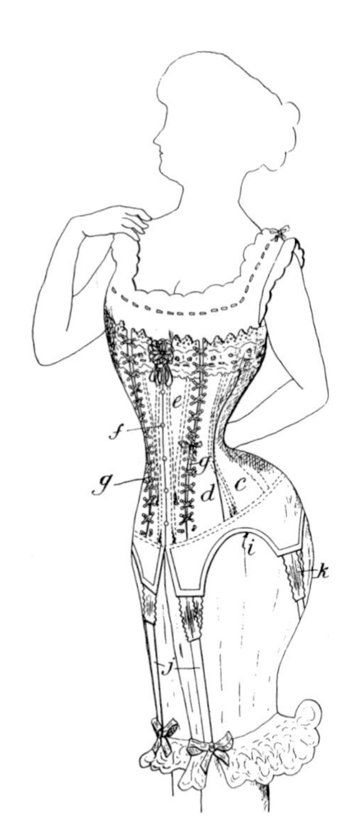

Even the style of drawing underwear reflected the self-confidence of the era. Maria Pressaud's patent diagram.

Brighton beach, seen here in 1859, had long been popular. (*Illustrated London News*)

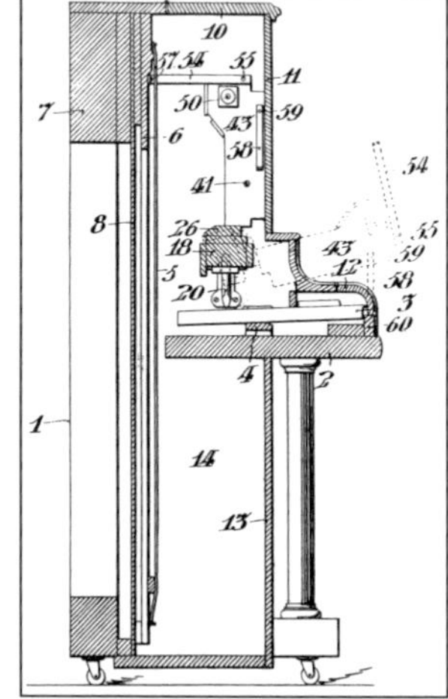

An automatic piano devised by Florence Heppe. Patent.

Hampson, was living on her own means in a large lodging house on Southport's fashionable Promenade. Here she invented her apparatus to practise roller-skating and 'rinking'.[13] Roller-skates, with four small wooden wheels, had become popular in the 1860s when a craze for them hit the United States and Europe. In 1876 Elizabeth Nash, a widow, and William Tytherleigh, an ironmonger, patented their roller skates.[14] Skates continued to be popular and Mrs Hampson's invention was for skilled skaters to practise 'complex evolutions in the more difficult and advanced art of what is known as figure skating'. A line was suspended between two poles from which a rope was hung and a pivoting cross bar attached to its lower end; from this two cords swung, on to which the skater grabbed, and the whole thing gave support for swirls and other complex movements to take place.

Mary Shippobottam, an entertainment contractor from Bolton in Lancashire, devised an intriguing roundabout in 1908.[15] Bicycles were placed on the outer edge of a roundabout and attached to its central pole with a long rod; as they were pedalled around enough power was generated to turn the whole thing. The surface of the roundabout was in four sections and, to increase the sense of fun and adventure, all the parts could be placed on the horizontal plane or any other angle; setting the quarters at different angles from each other would result in an undulating cycle ride, the speed of the roundabout dependent upon the amount of pedalling.

Sophia Barnacle's husband was headmaster of a school near Preston; clearly feeling the need to occupy their young charges she designed a type of helter skelter.[16] Instead of sitting on a mat, the rider used a chair that hung from a running wheel guided by a rail.

A 1905 Ascot Gown. (*Illustrated London News*)

Unlike previous methods that were powered by some sort of mechanism, hers relied on gravity to pull the rider along the rail and down a conical, spiral course from the top to the bottom of the tower.

The fashion for large head dresses, *c.* 1890s. (*Illustrated London News*)

In the two decades before the outbreak of war in 1914, there was an increase in patent applications by individual men and large companies for the machinery of battle: guns, rifles, firing mechanisms, army vehicles and munitions for ships. Generally, women are absent from this area, though Elizabeth Pennefather, a spinster from Bristol, did patent her method of 'Securing Cartridges in Cartridge-belts or Bandoliers' in 1901.[17] This was a leather belt around which thonging was wrapped and into which the cartridges were inserted. Apart from this there appear to be few, if any, inventions by women for the heavy machinery of warfare. They did, though, as has been seen, from the Crimean War onwards, concern themselves with the welfare of soldiers in the field, the safety of sailors at sea, nursing the sick and injured and rehabilitating the disabled. Over the years their designs have included stoves and food to feed the army, tents and stretchers for use in the field, bandages to cover wounds, furniture and devices for rehabilitation. Women's pacifist allegiances are well documented, but it is worth noting that in 1905 Baroness Sophie Felicita von Suttner was awarded the Nobel Peace Prize. She was born in Prague in 1843 and grew up to be appalled by the nationalistic militarism that surrounded her, making her an early advocate of the peace movement. Baroness Suttner wrote one book, *Lay Down your Arms*, to explain her philosophy for peace and became the Honorary President of the Permanent International Peace Bureau in Switzerland before being awarded the Nobel Prize.[18]

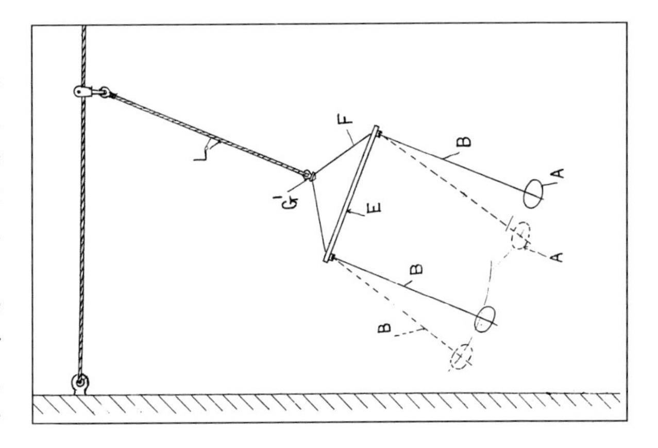

Skaters could practise their figure skating by grabbing on to Jane Hampson's suspended cords. Patent.

Women had proved they could solve problems, patent inventions, register designs, run companies, travel the world, organise suffrage movements, graduate

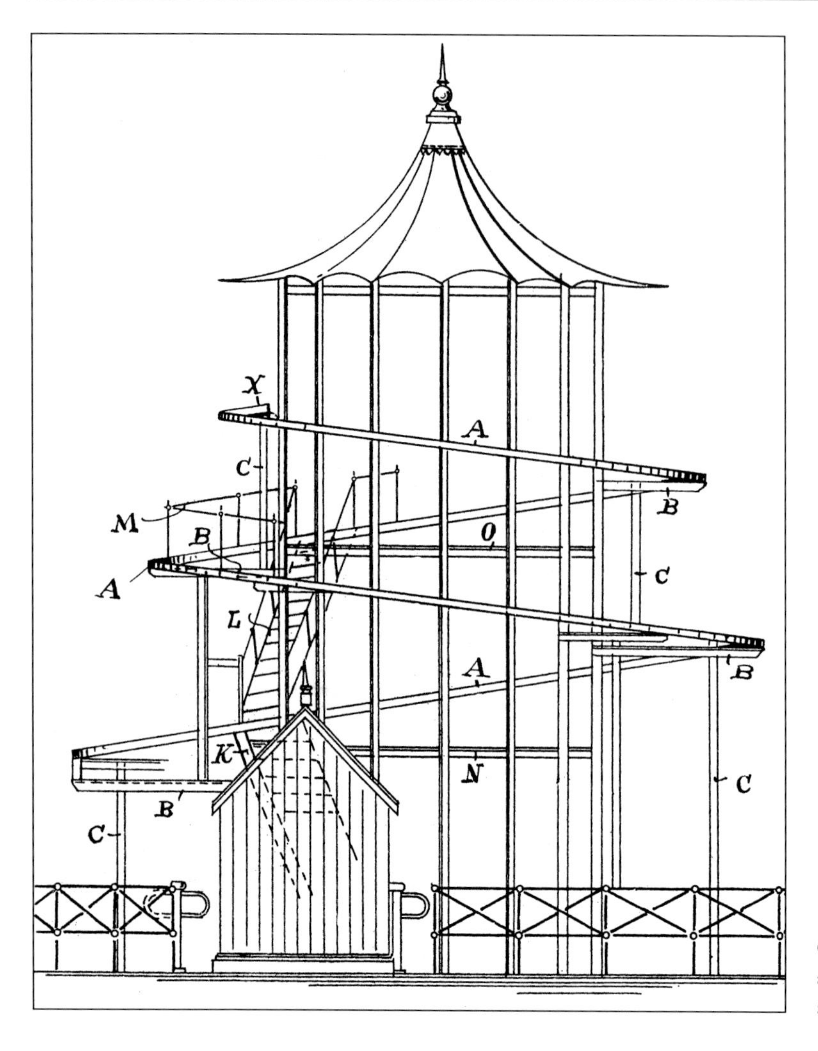

Chairs hung from the rail around Sophia Barnacle's helter skelter. Patent.

from courses in art, design, nursing and medicine, and exhibit and win awards at the trade exhibitions; nevertheless the worlds of industrial design, engineering and technology were still barred to them. Many of their inventions were practical in the context of the times in which they lived, while others may have been eccentric – but no more so than many of those devised by men. But the role of the inventor in a healthy, pluralistic society is to question, discover, ponder, imagine and design without the inhibition of the end result. As the industrialised world developed beyond the artisan craftsperson to mass manufacture, women were still expected to pursue the polite arts of needlework, floral decoration and arrangement of dinner parties for their husbands. It is no wonder many talented and able women started their own studios to be independent, and equally unsurprising that, without outside investment, they struggled.

Their sheer determination, against the odds; their ideas and imagination; the ability to problem-solve; and love of their work could with great benefit have been incorporated into the design, engineering and research departments of industry.

These women have left behind fascinating documents of their imagination, ideas, ingenuity and determination in the journey from a tincture of saffron via worming pills, straw bonnets, corsetry, anchovies, suspension bridges, divided skirts, ankle shields, life-saving signals, nursing appliances, geysers, furnaces, stoves, dishwashers and washing machines, to flying machines and fairground rides. Drawing on their immediate experience as well as situations further afield, these ingenious women have proved their ability to improvise, invent, adapt, devise and design to solve problems, despite encountering enormous obstacles. There are, though, the hundreds we will never know about, whose ingenious ideas were snapped up by husbands or employers to be produced as their own.

Baroness Sophie Felicita von Suttner, winner of the Nobel Peace Prize in 1905. (*The Nobel Foundation*)

Today little has changed in the worlds of art, design and science. An ingrained belief that it is not women's work, is hard to dispel. Few senior designers, architects, engineers, technicians and scientists in industry are women, even though many have been trained. It is disheartening to find many of the obstacles that these women had to conquer are still in place. In her recent report, Baroness Susan Greenfield, Director of the Royal Institution, was dismayed at prevalent attitudes to women in science which can be matched in all the professions – design and architecture, engineering, medicine, pure science and technology. Women still have to reverse intransigent attitudes.[19] These ingenious women have had to overcome that negativity, within themselves and others, in their firm belief that they had a good idea.

Their determination, imagination, vitality and passion, fundamental elements in any innovator, have been overwhelming. To have achieved what they did, when social attitudes and legislation were firmly pitted against them, is an expression of their individual self-confidence.

Appendix

Patents of Ingenious Women and General Chronology, 1637–1914

For chronological purposes the patents have been listed with the year of issue followed by the patent number, eg. 1850 GB1234. When ordering copies of British patents from libraries it is advisable to cite the number first followed by the year, eg. GB1234/1850. US patents are in numerical order without the year.

1625	**Accession of Charles I.**
1637	PatGB104. Amye Everard als Ball, Widowe. Ball's Patent. Preparing a Tincture of Saffron, Roses &c.
1642	**English Civil War.**
1649	**Execution of Charles I.**
1660	**Accession of Charles II.**
1678	PatGB204. Amy Potter, Widdow. Potter's Patent. An invention for makeing of Flanders Colbettine and all other laces of woollen to be vsed in and about the adorning or makeing-vpp of dresses and other things for the decent buriall of the dead.
1685	**Accession of James II.**
1688	**Accession of William III and Mary II.**
1702	**Accession of Queen Anne.**
1714	**Accession of George I.**
1715	PatGB401. Thomas Masters of Pensilvania, Planter. A new Invencon found out by Sybilla his wife, for Cleaning and Curing the Indian Corn Growing in the severall Colonies in America.
1716	PatGB403. Sybilla Masters, wife of Thomas Masters, Merchant, of Pensilvania. A new way of Working and Staining Straw, and the Platt and Leaf of the Palmeta Tree, and Covering and Adorning Hatts and bonnetts, in such a manner as was never before done or Practised in England or any of our Plantacons.
1731	PatGB532. Elizabeth Coppin, Widdow. Fluxing and Fixing Mundic.
1737	PatGB560. Jane Vanef, widow and hoop petticoat maker of the Parish of Saint Anne, in the Liberty of Westminster, in the County of Middlesex. Hoops for Petticoats.
1760	**Accession of George II.**
1769	Mrs Eleanor Coade opens architectural mouldings factory, Lambeth, south London.

1775	**American War of Independence.**
1787	**Designing and Printing of Linen Act.**
1789	**French Revolution.**
1792	Mary Wollstonecraft, *A Vindication of the Rights of Women* published.
1799	**Start of Napoleonic Wars.**
1799	PatGB2365. Ann Wilcox and Edmund Ludlow, City of London. Intire New or Improved Playing Cards, to be named 'Brilliant new Invented Knight's Cards'.
1800	Pat GB2457. Martha Gibbon, Dressmaker, King Street, Covent Garden, County of Middx. A certain new stay for women and others.
1801	PatGB2485. Ann Young, St James Square, Edinburgh. New invented apparatus, consisting of an oblong square box, which, when opened, presents two faces or tables, various dice, pins, counters &c. contained within the same, by means of which six different games may be played.
1802	Madame Marie Tussaud arrives in London with her collection of waxwork replica figures.
1803	PatGB2702/2703. Elizabeth Bell, Hampstead. An Artificial Method of Sweeping Chimnies and of constructing them in such a manner as to lessen danger and inconvenience from fire and smoke.
1807	PatGB3019. Elizabeth Bell, Spinster of Blackheath. Certain Improvements in an artificial method of sweeping chimnies.
1808	PatGB3129. Rebecca Ching, Rush Common, Lambeth. Certain Improvements in a Medicine then called 'Ching's Worm Destroying Lozenges' (Ref. Pat.1796/GB2121).
1809	Mary Kies, Connecticut, USA. Weaving Straw into Silk or Thread. Patent destroyed.
1809	PatGB3221. Elizabeth Perryman, Greek Street, Soho. An Improved Street and Hall Lamp and the necessary apparatus for expediting the trimming, lighting and cleansing the same.
1811	PatGB3405. Sarah Guppy, wife of Samuel Guppy, Merchant, Bristol. A New Mode of Constructing and Erecting Bridges and Railroads without Arches or Sterling, whereby the Danger of their being washed away by Floods is Avoided.
1812	PatGB3549. Sarah Guppy, wife of Samuel Guppy, Merchant, Bristol. Certain Improvements in Tea and Coffee Urns.
1815	PatGB3893. Elizabeth Beveridge, Hatton Garden, City of London. Improved Bedstead.
1815	PatGB3930. Grace Elizabeth Service. A New Method of Manufacture of Straw with Gauze, Nett Webb and other similar articles, for the purpose of making into hats, bonnets, workboxes, workbags, toilet boxes and other articles.
1815	**Duke of Wellington defeats Napoleon at Waterloo.**
1819	PatGB4370. Sarah Thomson of Boyd's Rope Walk, Rotherhithe, Surry. Widow of Archibald Thomson, late of Church St, Blackfriars. 'A Machine for Cutting Corks communicated to me by my late husband . . . and also by my late son, Alexander Thomson, also deceased.'
1819	**Factory Act: children under nine not allowed to work in cotton mills.**
1820	**Accession of George IV.**
1824	Society of Arts Premium of 15 guineas. Lucy Hollowell of Neithrop, nr Banbury. Leghorn Straw . . . using seeds imported by the Society from Connecticut . . . to produce superior straw/grass.
1825	Society of Arts Silver Ceres Medal. Mrs Lowrey of Exeter (see Pat1829/GB5630). A bonnet made of double wheat straw.

1825 Society of Arts Rewards for Bonnets and Hats made of British materials, platted and knit in imitation of those imported from Leghorn: Sophia Dyer and Anne Dyer of West Meon, Alton, Hants. Mrs Venn and Anne Venn, Hadleigh, Suffolk. Lucy Hollowell, Banbury. The children of Mrs Villebois's school at Adbury, Berks. Mary Marshall.

1826 Society of Arts Large Silver Medal Presented to Miss Petner for (silk) worms grown in England.

1828 PatGB5630. Jane Bentley Lowrey, wife of Thomas Sampson Lowrey of Exeter. Straw Hat Manufacturer. Certain Improvements in the Manufacture of Hats and Bonnets.

1829 PatGB5809. Margaret Knowles, Spinster, Lavender Hill, Battersea, Surrey. An Improvement in Axletrees for and Mode of Applying the same to Carriages.

1830 PatGB6057. Marie Elizabeth Antoinette Perths, Spinster. Rue de Bai, Paris. The Fabrication of a Coal fitted for the Refining and Purifying of Sugar and other matter, and to restore the coal which has served for that purpose.

1830 Accession of William IV.

1831 PatGB6186. Sarah Guppy, Widow, Clifton, Bristol. A Method of Applying and Arranging Certain Articles, Parts or Pieces of Cabinet Work, Upholstery, and other Articles commonly or frequently applied to bedsteads and hangings, and also others not hitherto so applied.

1833 Society of Arts Silver Ceres Medal – Mrs Davies Gilbert of Eastbourne. Method of bringing seaside shingle into cultivation, and of making water tanks.

1833 Factory Act – in textile mills – children nine to twelve years to work maximum 48 hours per week, thirteen to nineteen year olds a 69-hour maximum week, those under nine not allowed to work.

1837 Accession of Queen Victoria.

Sound transmitted along a cable attached to the railway line from Euston to Camden Town.

1838 PatGB7640. Jean Francis Isidore Caplin, Portland Street, Middx (his wife, Roxey Caplin, not included in patent application). Improvements in Stays or Corsets, and other parts of the dress where lacing is employed, and in instruments for measuring for corsets or stays, and for the bodies of dresses.

1838 London to Birmingham Railway, Euston Station and Great Western Railway opened.

1839 PatGB8086. Josephine Julie Besnier de Bligny, Rue de Condé, Paris and now residing at the Commercial Hotel, Leicester Square, Middx. Improvements in Umbrellas and Parasols.

1839 PatGB8234. Francis Gibbon Spilsbury, chemist, of Walsall. Marie Françoise Catherine Doetzer Corbaux, Artist, of Upper Norton Street, in the County of Middlesex. Alexander Samuel Byrne, Gentleman of Montague Square in the County of Middlesex. Preparing and Applying Paints or Pigments.

1840 Ada, Countess Lovelace, works on the language for Charles Babbage's analytical engine.

1842 Factory Act – (textile factories) women and girls from thirteen to eighteen to work maximum 12 hours per day. Under thirteen year olds to work maximum 6½ hours per day and 3 hours education.

***Enquiry into the Sanitary Conditions of the Labouring Population*, by Edwin Chadwick.**

1843 PatGB9740. Sarah Beadon, Hope Corner, Taunton. Improvements in apparatus for regulating the inclination of vessels for the purpose of drawing off liquids contained therein.

1843 Society of Arts Mechanics Award. Mrs Jemima Goode, Ryde, Isle of Wight. Improved Gothic Window Blind.

1843 PatGB9900. Margaret Marshall, Manchester, County of Lancaster. Plastic Composition.

1844 PatGB10267. Sarah Coote (formerly Sarah Guppy), Clifton, Bristol. Improvements in Caulking Ships and other vessels.

1844 PatGB10281. Elizabeth Cottam, wife of George Cottam, Winsley Street, Oxford Street, Middx. Improvements in Heating what are called Italian Irons.

1844 Ragged Schools opened by Lord Shaftesbury.

1844 Society of Arts Silver Isis Medal. Mrs T. Allan, 14 Hart Street, Bloomsbury. For the introduction of bees to New Zealand.

1845 PatGB10572. Anna Maria Stowell, straw bonnet manufacturer, Gloucester Place, Islington Green, Middx. Thomas Little, manufacturer, Hoxton Old Town. Improvements in the Manufacture of Ladies' Bonnets or Hats.

1845 Thomas Cook organises railway journeys for passengers.

1847 PatGB11549. Elizabeth Lutel on behalf of M. César Luc Louis Oudinot of Paris. Producing a Certain Texture Elastic in Same.

1848 PatGB12075. Elizabeth Wallace, spinster, Laurel Lodge, Cheltenham. Certain Improvements in Facing, Figuring, Designating, Planning, and otherwise Fitting up Houses and other Buildings, parts of which are applicable to articles of furniture.

1848 PatGB2198. Elizabeth Dakin, Widow, St Paul's Church Yard, City of London. Improvements in Cleaning and Roasting of Coffee, in the apparatus and machinery to be used therein, and also in the apparatus for making infusions and decoctions of coffee.

1849 First Woman Doctor in the USA, Elizabeth Blackwell.

1849 PatGB12607. Charlotte Smith. Wearing Apparel: stays and corsets.

1849 Society of Arts Prizes presented by HRH Prince Albert, President of the Society of Arts. Miss Stanley for her exertions in establishing schools for instructing the children of the poor in the manufactures of lace in Norwich.

1850 Society of Arts Large Silver Medal. Miss Mary Ann Shackleton for a series of original drawings of British wild flowers from nature.

1850 Society of Arts Silver Isis Medal. Mrs Machlauchlan for needlework.

1850 Society of Arts Silver Isis Medal. Miss Catherine Marsh for a series of original drawings of wild flowers from nature.

1850 Factory Act – women and young people allowed to work between 6 a.m. and 6 p.m. with meal break.

1851 The Great Exhibition, London.

Some of the Exhibitors at the Great Exhibition:

The Dowager Lady Juliana Ashburnham, Producer, Hastings. A bag of hops, grown within three miles of the sea in the parish of Guestling, Sussex.

Mme Roxey Caplin, Manufacturer, Designer and Inventor, Berners Street, London W. Awarded the Prize Medal for self adjusting corsets, child's bodices, ladies' belts.

Liza Davy, Inventor and Manufacturer, Grosvenor Street. New registered riding stays, nursing stays, dress stays, and stays of the usual kind.

Mary Galton, Designer and Manufacturer, Fitzroy Square and Pentonville. Sofrano standard rose tree, mignonette. Modelled in wax.

Maria Gray, Designer and Inventor, Hoxton. Group of flowers made of human hair.
Mary Mathews, Inventor, Westbourne Street, Hyde Park Gardens. 'Astrorama'.
Matilda Pullan, Designer, Inventor and Manufacturer, Regents Park, London. Modern point lace.
Anne Rayner, Designer, Berners Street, London W. Specimens of diamond engraving upon black marble.
Charlotte Readhouse, Newark on Trent. A Lunar Globe.
Sophia Seltzer, Inventor, Designer and Producer, Upper Ranelagh Street, Pimlico. Chair for spinal curvature.

1851 **Baron von Reuter's News Agency.**

1852 PatGB1064. Jean François Isidore Caplin, Orthorachidist (his wife Roxey Caplin, not included in patent). Strawberry Hill, Near Manchester. Apparatus for Preventing or Curing a Stooping of the Head or Body.

1852 PatGB1062. Susan Walker of Hersham, Surrey. Improvements in Clogs and Pattens.

1852 **Patent Act (Design and Copyright).**

1853 PatGB554. Mary Ann Smith, of Wimpole Street, Mary-le-bone in the County of Middlesex. Improvements in the Manufacture of Toys, Models, and other like articles or ornament or utility.

1853 PatGB2368. Mary Ann Davy and Ann Taylor, Islington, Middlesex. Improvements in the Mechanical Application of Brushes.

1853 **Factory Act – children to work only from 6 a.m. to 6 p.m. with 1½ hours meal break.**
Smoke Abatement Act for the Metropolitan Area.

1853–6 **Crimean War.**

1854 PatGB145. Marie Louise Lise Beaudeloux, Paris, Empire of France. A Self Acting Cradle, with Improved Mattress.
PatGB146. A Candle Stick Working by Machinery.

1854 PatGB1913. Marie Louise Lindheim, Independent Lady, Paris. Certain Improvements in the Manufacture of Bonnets or Caps.

1854 PatGB2659. Maria Morrison, Chelsea, Middx. A Mode for Preserving Inscriptions and Paintings on Glass, applicable for monumental and other tablets.

1854 PatGB2194 (Prot. Refused see 1855/790). Louisa Monzani, Widow and Administratrix, Greyhound Road, Old Kent Road, Surrey. Improvements in the Manufacture of Folding Chairs.
PatGB2202. Bedsteads, packing cases for the same.
PatGB2203. Brushes and Brooms.

1854 PatGB273. Margaret Williams, Chelsea, Middx. Improvements in Suspending Swing Looking or Dressing Glasses.

1854 Florence Nightingale, Mary Seacole and others nurse in the Crimea.

1855 PatGB726. Elizabeth Abbott and Matilda Abbott, Gentlewomen, Horningsea, Cambridge. Improvement in Stays.

1855 PatGB790. Louisa Monzani, Greyhound Place, Old Kent Road, Surrey. Improvements in Folding Stools and Folding Chairs.

1855 PatGB1267. Mary Staite, Liscard in the County of Chester. The Manufacture of a New Black Paint.

1855 PatGB2837. Agnes Wallace of Nether Place Bleach Works, in the County of Renfrew, North Britain, and John Wallace of the same place, Bleachers. Improvements in Bleaching, Washing, or Cleansing Textile Fabrics and Materials.

1856 PatGB7. John Thurell, Castle Street East, Oxford Street, Elizabeth Mary Muller, 58 Greek Street, Soho, Middx and John Robert Chidley, Gresham Street, City of London. Improvements in Transmitting Facsimile Copies of writings and Drawings by means of Electric Currents.

1856 PatGB791. Frances Young, Norwich, Norfolk. An Improved two-wheeled open vehicle or carriage.

1856 PatGB1791. William Griffin and Elizabeth Duley of Northampton. Improvements in studs and buttons for fastening articles of dress.

1856 PatGB2330. Maria Farina, Hanway Street, Oxford Street, Middx. An Improved Tooth Powder.

1857 Matrimonial Causes Act.

1857 PatGB1410. Maria Bounsall Rowland, Widow, Acton Green, Middx. Improvements in Soap and Detergent Preparations or Compounds.

1857 PatGB2420. Charlotte Delevante, Spinster. Improvements in Bouquet Holders.

1858 PatGB873. Maria Ross, Gallowtree Gate, Leicester. Improvements in the Manufacture of Frames for Looking Glasses, Pictures and other representations.

1858 PatGB2749. Ann Elizabeth Davis, Pulboro' Place, Harleyford Road, Vauxhall. Richard Wright of 28 Grosvenor Park, Camberwell. Manufacture of Colouring Matter for Spirits etc.

1859 PatGB2. Helen Catlin Traphagen, New York City, USA. Improvements in Skirts for Ladies.

1859 PatGB763. Elizabeth Steane, Widow, Manor Rise, Brixton. An Improved Means or Apparatus for Preventing Candles Dropping or Guttering.

1859 PatGB873. US23536 and US115935. Martha J. Coston, Washington DC, USA. Improvement in the Construction and use of Fireworks for Signals/Pyrotechnic Night Signals.

1859 PatGB1014. Charlotte Mansel, Plymouth. A Folding Travelling Case (see 1870/2713).

1859 PatGB1778. Elizabeth Merrell, tin and zinc plate worker, Little George Street, Minories, City of London. Improvements in Apparatus for washing and cleansing.

1859 PatGB2573. Elizabeth Ann Sellon Burgess, 107 The Strand, Middx. Improvements in the Preparation of Anchovies.

1860 PatGB656. Marie Joséphine Elisabeth Julienne, Boulevard St Martin, Paris. An Improved Belt to be Applied in the Bathing Vessels and in Electrical Apparatus Connected Herein.

1860 PatGB1952. Emma Benjamin Orange, Barentin in the Dept of Seine the Inferieure in the Empire of France. An Improved Method of an Apparatus for Unloosening Horses Instantly from a Carriage when Frightened or Running Away.

1860 PatGB2161. Madam Louise Fauve, 42 Rue Laffite, Paris. An Improved Method of Preserving Various Alimentary Substances.

1861 PatGB1348. Frances Ann Whitehead, Whitehead's Grove, Chelsea, Middx. Improvements in Treating Cream or Milk and in Obtaining Butter and other Products Therefrom.

1861 PatGB2367. Elizabeth Steane, Commercial Place, Brixton Road, Kent. Improvements in Apparatus, by the use of which the Dropping or Guttering of Candles is Prevented.

1861 PatGB2740. Elizabeth Ann Malling, Spinster, Whitehead's Grove, Chelsea. Improvements in Glass Cases for the Cultivation of Plants and Flowers.

1861 PatGB3059. Charlotte Craddock, Dressmaker, Orchard Terrace, Kensington. An Improved System or Method of Cutting out Ladies' Dresses.

1861	PatGB3216/18. Charlotte Smith, Wife of Jabez Smith of Bedford. Improvement in Stays.
1862	PatGB2464. Emma Louise Duncan. Inverness Road, Bayswater, Middx. Splints.
1862	PatGB2712. Mary Anne Beale, Barnsbury, Middx and John Beale, Gentleman, Maidstone, Kent. Improvements in the Preparation or Manufacture of Manure.
1863	PatGB1082. Margaret Barland, Mount Street, Grosvenor Square and Edward Henry Cradock Monckton, The Cavendish Club, Regent Street, Middx. and Thomas Barland, Eau Clair, Wisconsin, USA. Improvements in Apparatus for Withdrawing Milk from Cows and other Mammiferae and for Conducting it when Withdrawn to Appropriate Receivers.
1863	PatGB1646. Camille, Baroness de Lavenant, Paris. Certain Compositions for Protecting Metals and Metallic Articles from oxidation, and for coating slate, bricks, pottery and ceramic ware.
1863	PatGB1788. Agatha Montleart, Widow, of Mildmay Park, Middx and William Tent of City of London. An Improved Mode of Attaching Hooks to Furniture or Fabrics for Suspending Dresses or parts to dresses, fabrics, curtains and other articles of upholstery or apparel.
1863	PatGB1794. Anne Catherine Veuve Durst-Wilde of Paris. Producing Raised Patterns on Hats, Caps and Bonnets.
	PatGB2804. A.C. Durst-Wilde, Widow, 119 Cheapside in the City of London and Rue de Caire, Paris. Ornamenting Hats and Bonnets.
1863	Prov.PatGB1948. Catherine or Katy Liddle, Cook and Housekeeper, Comely Bank, near Edinburgh, North Britain. An Improved Method of Cooking an Egg, and an Improved Pan or Apparatus for the same.
1863	PatGB2168. Elizabeth Collier, Milliner and Crinoline Manufacturer, High Street, Leicester. Improvements in Crinolines and Crinoline Fastenings.
1863	PatGB2613. Mary Anne Borle, Kensington, Middx. Improvements in Portable Writing Cases and Despatch Boxes.
1863	PatGB3119. Hon. Susan Tuchet, Spinster, 14 Liverpool Street, Dover. An Improved Roller for Window Blinds.
1864	PatGB396. Jane Elizabeth Tuchet, Spinster, Liverpool Street, Dover. Improvements in Apparatus for Cleaning Windows.
1864	PatGB1449. Hon Susan Tuchet. Improvements in the Construction of Rollers for Window Blinds.
1864	PatGB582. Madame Jobert of Dole, France. Improvements in the Manufacture of Washing Blues.
1864	PatGB1880. Elizabeth Brimson, Spinster, Frome, Somerset. Improvements in Envelopes or Covers for Bottles or Jars.
1864	PatGB2117. Elizabeth John, Widow, Nottingham. Improvements in the Construction of Bedsteads.
1864	PatGB2301. Amelia Higgins, Freeschool Street, Parish of St John's, Horselydown, Southwark. Improvements in Embroidering or Ornamenting Woven or Spun Fabrics, applicable to articles of wearing apparel, ornamental decorations of furniture and other similar purposes.
1864	Octavia Hill. Housing project for the poor in Marylebone.
1864	**Factory Act – health and safety regulations for dangerous industries.**
1865	PatGB1628. Sarah Martha Buckwell, Luino, Italy. Improvements in the Method of and Apparatus for Effecting and Recording Telegraphic Communications.
1865	PatGB 2430. Jane Elizabeth Tuchet, Spinster, Liverpool Street, Dover. Securing or Fastening Envelopes.

1865 PatGB3298. Marie Louise Changeur, Boulevard Beaumarchais, Paris. An Improved Kind of Clasp or Dress Preserver to prevent ladies' dresses from trailing along the ground.

1865 Movement for Women's Suffrage begins.
Telegraphic cable laid from England to India.

1866 PatGB589. Charlotte Elizabeth Treadwin. Improvements in the Manufacture of Certain Kinds of Lace.

1866 PatGB1153. Rachel Stackhouse, Widow, Central Hill Cottage, Central Hill, Norwood, Surrey. Improvements in Charts, Maps, Plans Illuminated and other Printed Articles by the use of a waterproof and untearable material as a substitute for paper.

1866 PatGB1321. Desirée Gautier, New Cavendish St, Portland Place, Middx and Albert Domier, City of London. Improvements in Apparatus for Supporting and Extending the Skirts of Ladies Dresses.

1866 PatGB1446. Clara Löwenberg, New York City, USA. An Improved Composition for Beautifying the Complexion.

1866 PatGB2742. Elizabeth Mignot, Rue de Tracy, Paris in the Empire of France. An Improved Invention for Distending the Lower Part of Ladies' Dresses or Petticoats.

1866 Elizabeth Garrett Anderson (first female doctor in England), appointed medical attendant and establishes New Hospital for Women.
Transatlantic telegraphic cable laid.

1867 PatGB443. Mlle Marie Louise Changeur, Boulevard Beaumarchais, Paris. Improvements in Corsets.

1867 PatGB481. Louisa Hartland Mahon, Bayswater. A New or Improved Method of Working or Producing Designs in Wool.

1867 PatGB592. Anne Catherine Laurys, Merchant, Louvain in the Kingdom of Belgium. Elastic Fabric. A New and Useful Fabric for Elastic Stocking, Stays or Corsets, Bandages and other similar and analogous articles.

1867 PatGB1546. Louisa Slatter, Drayton, Oxford. Improvements in Cases for Packing and Transporting Butter and other dairy produce.

1867 PatGB1792. Ann Littlejohns Panter, Pakenham Street, Grays Inn Road, London. An Improved Seat for Babies or Young Children.

1867 PatGB1951. Elizabeth Martha Brown, Bury St Edmunds, Suffolk. Improvements in Apparatus for Carrying Umbrellas.

1867 PatGB2297. Caroline Holograph, Widow, Brunswick, Germany. Improvements in Apparatus to be Applied to Chimney Tops.

1867 PatGB2358. Rebecca Joseph, Corset Manufacturer, Westminster Road, Surrey. Improvements in Stays or Corsets.

1867 PatGB2482. Henry Octavius William Cooper and Elizabeth Foster Cooper, St George's Road, Hanover Square, London. A Liquid or Composition (of a deodorizing nature) Mixed with Water, whereby roads, streets, and other places are kept moist and free from dust for several days after once watering.

1868 PatGB16. Bertha Vette, wife of Theodore Vette, Berlin, Prussia. Improvements in the Manufacture of 'crossovers', Scarves and other articles of apparel known as frame and knitted woollen goods.

1868 PatGB26. Mary Eliza Roy of Upper Norwood and Louisa Prevett of Penge, Surrey. An Improved Receptacle for Containing Needles, Thread and other articles.

1868 PatGB361. Maria Alice Wilson, Victoria Road, Kensington, Middx. Improvements in Spring and other Mattresses, Couches and Cushions.

1868 PatGB609. Sarah Elizabeth Saul, New York City, USA. Improvements in Apparatus Employed for Cooking, Boiling, Melting and Evaporating Purposes.

1868 PatGB1602. Charles Seely, Charles Eames, William Clark and Mary Louisa Booth of City and State of New York, USA. An Improved Mode of Embalming or Preserving Dead Bodies.

1868 PatGB1609. Sarah Saul, New York City, USA. Improvements in Appliances for Cooking.

1868 PatGB1658. Caroline Garcin and Amelia Garcin, Colmar, France. Improvements in Sewing Machines.

1868 PatGB1664. Mary Ann Duffy, New York City, USA. An Improved Device for Marking and Creasing Tucks in Cloth or other material upon a sewing machine.

1868 PatGB2877. Henrietta Vansittart, Richmond, Surrey. Improvements in the Construction of Screw Propellers.

1868 First School of Nursing – St Thomas's Hospital, London.

1869 PatGB713. Antoinette Vidal, Paris, France. A New Sort of Porcelain and a New and Improved Process for Manufacturing the Same.

1869 PatGB778. Elizabeth Woolcott Slade and Maximelia Slade of Wilton near Salisbury. An Improved Portable Oven.

1869 PatGB 2073. Santoni Faltorini, Madame la Marquise de Livry, née Louise Appoline Oudot de Dainville, Brussels, Belgium. An Improved Mode of Measuring or Dividing Time, applicable to chronometers, sextants and other such like instruments.

1869 PatGB2404. Jane Cross, Farnworth, Manchester and John McCann of Lancaster. Improvements in the Manufacture of Dry Soap for Washing, Cleansing, Calendering and Finishing Cotton and Cotton Fabrics, Wool and Woollen Fabrics and Mixtures of the same.

1869 PatGB3092. Mrs Hannah G. Suplee, San Francisco, California, USA. Improvements in Sewing Machine Needles.

1869 Octavia Hill with Edward Denison. Charity Organisation Society to investigate the living conditions of the poor.

1870 PatGB182. Mrs Mary Carpenter, San Francisco, California, USA. Improvements in Needles and Needle Arms for Sewing Machines.

1870 PatGB2332. Elisabeth Hafner de Bodmer, Paris. Improvements in the Construction of Carriage Wheels.

1870 PatGB2713. Charlotte Mansel, Spinster, Haverstock Hill, Middx. An Improved Knitting Sheath.

1870 PatGB2662. Elizabeth A. Clarke, Chicago, Illinois, USA. Improvements in Bows, Scarves and Similar articles to be worn with Collars.

1870 PatGB2822. Sarah Beauchamp Bush, Upper Westbourne Terrace, London W. Improved Means of and Apparatus for Preventing the Entrance of Flies and other Insects and Reptiles into Rooms or Apartments of Buildings, Ships and Carriages.

1870 PatUS98959. Mary Hanson, Manston, Wisconsin, USA. Medical Compound.

1870 PatUS101321. Anna Smith, Pittsburgh, USA. Washing Machine.

1870 PatUS101576. Mary D. Brine, Chicago, USA. Portable Swing.

1870 PatUS101934. Linda Spigelmyer, Hartleton, Pennsylvania, USA. Child's Body Brace and Supporter.

1870 PatUS102338. Annie Vogel, New York and Fannie Krebs, Georgetown DC, USA. Hair Curling Pin.

1870 PatUS102868. Amanda Rosbrugh, Panora, Iowa, USA. Bleaching Straw Goods.

1870	**Married Women's Property Act – women able to keep £200 of own earnings.**
	Elementary Education Act for girls and boys.
1871	PatUS115935. Martha Coston, Washington DC, USA. Night Signals.
1872	PatGB1198. Sarah Johnson and Alfred Johnson, Leicester. An Improved Plumb Rule and Level.
1873	PatUS139580/140508. Amanda Jones, Inventor, Clinton, Wisconsin, USA. Vacuum Fruit Jars.
1874	PatGB607. Amelia Louisa Freund, trading as Amelia Lewis, Authoress, Editor and Publisher. Southampton Street, Middx. An Improved System or Method of Domestic Cooking and Improved Apparatus and Utensils to be employed therein.
	PatGB2849. Amelia Freund. A New or Improved System or Method of Domestic Cooking and New or Improved Apparatus or Utensils to be employed therein.
1874	**Sophia Jex-Blake. First Medical School for Women.**
1874	**Factory Act – minimum working age nine years, women and young people to work maximum ten hours per day in textile mills.**
1875	PatGB545. William Tytherleigh, Ironmonger and Builder, Upper Weymouth, St Marylebone. Elizabeth Nash, Widow, of 30 Weymouth Street, London. Improvements in the Construction of Roller.
1875	PatGB2936. Catherine Richardson, City of London. Flat Pleating or Kilting Machines.
1875	PatGB3558. Emily Stewart Kerr, Ladbroke Grove. An Improved Device for Lifting and Holding up the Skirts of Ladies' Dresses.
1875	PatGB4845. Elizabeth Spink of York. Improved Combination of Ingredients or Compounds for Curative Purposes.
1876	PatGB1010. Sullivan Haynes Penley, Rialdo Dorman and Annie Judson Dorman of the USA. Improvements Relating to the Utilization of Rattan Pith for the Manufacture of Veneers for the Construction of Carriage Bodies and for other like purposes.
1876	PatGB1727. Elizabeth Clark, 40 Conduit Street, London. Improvements in Abdominal Belts and Trunks Attached.
	Pat.GB3027. Elizabeth Clark. An Improved Combined Under dress for Ladies (Chemise).
1876	PatGB3216. Maria Procopé, Stockholm, Sweden. Apparatus for Tuning Organs.
1876	PatGB3487. Emma Elizabeth Ashing, Camden Town, County of Middlesex. Improvements in Solitaires and Studs and other Similar Dress Fastenings.
1876	PatGB3585. Amelia Pitts Armstrong, Electrician, and others, City and County of New York, USA. Ventilating, Disinfecting and Fog Alarm Apparatus for Ships, etc.
1876	PatGB4120. Ada Sarah Ware, St Martin's Lane, London. An Improved Horse Rein Holder.
1876	**Centennial Exhibition, Philadelphia. Women's Pavilion.**
1877	PatGB345. Charlotte Crater, Marine Parade, Folkestone. Apparatus for the Propulsion of Vehicles.
	PatGB389. Charlotte Crater. Signals.
1877	PatGB3128 and 3129, US192522 and 193012. Mary Boyd Cummings, Washington DC, USA. Improvements in Postal Envelopes and Postal Cards.
1877	**Married Women's Property Act – Scotland.**
1878	PatGB68. Emma Juliana Gammon, the Firs, Leyton, Essex. Improvement in Stays.

1878 PatGB330. Elizabeth Frances Kelly, All Saints Road, London W. An Improved Musical Instrument.

1878 PatGB1051. Amelia Louisa Freund, Trading under the name of Amelia Lewis of Regent's Square, London. Improvements in Cooking Stoves and Utensils.

PatGB1716. Amelia Louisa Freund, Oldham. Improvements in Cooking Stoves and Utensils Specially Adapted for Military and like uses.

1878 PatGB1350. Elizabeth Marriott, Northampton. An Improved Washing or Cleansing Compound.

1878 PatGB1397. Mary Permellia Carpenter, New York, USA. Improvements in Sewing Machines, Chiefly Designed for Straw Braid Work.

1878 PatGB1574. Elisabeth Loeper, Magdeburg, Germany. Button Working Machines.

1878 PatGB1910. Mary Brewster, Logie Green, Edinburgh, Midlothian. Improvements in the Manufacture of Feather Fabric Adapted for Use in the Production of Articles of Dress and for other Purposes.

1878 PatGB2213. Elizabeth Jane Corbett, Physician, San Francisco, California, USA. Improvements in Devices for Removing Gases from Sewers and Destroying them by Heat.

1878 PatGB2490. Mary Pip, Porchester Road, Bayswater. Improvements in Cork Soles or Lining for Boots and Shoes.

1878 PatGB3118. Sarah Jane Blick, Lansdowne Works in the City of Worcester. Improvements in Expanding Canopies Applicable to Carriages, Tents and other like purposes.

1878 Factory and Workshops Acts – state of workshops.

1879 PatGB2988. Mary Jane Cooke, Heaton Moore, Near Manchester. Improvements in Cesspools, and in Apparatus or Arrangements for Filtering or Separating the Matters Flowing Thereinto.

1880 PatGB334. Mary Innes, Manchester. Improvements in the Construction of Fasteners for Elastic Bands or Garters.

1880 PatGB433. Susan Hibbard, Geneva Lake, County of Walworth, Wisconsin, USA. Improvements in the Manufacture of Feather Dusters.

1880 PatGB876. Madame Courmont, Widow, née Marie Bérenger, Lyons, France. Improvements in the Manufacture of Openwork Fabrics.

1880 PatGB1595. Madame Adele Engstrom, Paris. Improvements in Curry-combs.

1880 PatGB1283. Elizabeth Langdon. Improvements to Stays and Corsets.

1880 PatGB1601. Maria Beasly, Philadelphia, Pennsylvania, USA. Improvements in Life Saving Rafts.

1880 PatGB4151. Elizabeth Allibert, Brompton Square, Middx. Improvements in Stays and Corsets.

1880 PatGB2280. May Arnold, Spinster, Acton, Middx. Incubating Apparatus.

1880 PatGB4657. Elizabeth Eavestaff, Milliner, Upper Berkeley Street, Portman Square, Middx. Expanding Dress Stand.

1880 PatGB2797. Barbara Faucher and others, New Jersey, USA. Machines for Cutting Screw Threads on Pipes and Couplings.

1880 PatGB1872. Mary Tyler Foote, Boston, Massachusetts, USA. Improvements in Game Apparatus.

1880 PatGB1918. Clara Rouse, Chislehurst, Kent. Sootscreen for use in Cleaning Chimneys.

1880 PatGB2545. Annie Eliza Scot, Queensgate, South Kensington. A New or Improved Visitors' Tablet.

1880 PatGB2760. Elizabeth White Stiles, Hartford, Connecticut, USA. Improvements in Suspended and Self Levelling Receptacle.

1880 PatGB3069. Eliza Priestley. Ventilating Rooms and Buildings.

1880 PatGB3674. Harrietta Röhner and Louis Pagel, Liverpool, County of Lancaster. Manufacture of Stockings.

1880 PatGB4220. Harriet Hosmer, Rome, Kingdom of Italy. Apparatus for Obtaining Motive-power.

1880 Margarete Steiff establishes mail order company selling soft toys (Steiff bear, 1903).

1882 PatGB631. Alice Wardrop, Stay Maker, Hanover Street, Middx. Improvements in or Additions to Corsets, Stays or Body Belts.

1882 PatGB872. Clara Brinkerhoff and George Cuming, New York City, USA. Improvements in and Appertaining to Electrodes for Telegraphic Instruments and other Electrical Instruments.

1882 PatGB1227. Emma Loxton, Horbury, Wakefield, County of York. Paste or Cream for Washing Linen, Flannel etc.

1882 PatGB1308. May Arnold, Spinster, Acton, Middx. Artificial Hatching Apparatus.

1882 PatGB1751. Anna Dormitzer, New York, USA. Improvements in Apparatus Applicable to Window Cleaning, Chairs or Fire Escapes.

1882 PatGB2949. Emily Smyth, Shepherds Bush, Middx and Martha Wood, Stafford. Improvements in Boxes made of Cardboard or similar material and in the method of manufacture of the same.

1882 PatGB5486. Clara Warner, South Place, Finsbury, Middx. An Improved Process for the Manufacture of a Compound to be used in Rendering Animals and Vegetable Fibres and Fabrics Water Repellent and Proof against Damage from Moths, Mildew and Vermin.

1882 PatGB4589. Catherine Adams, Asheville, County of Buncombe, North Carolina, USA. Improvements in Corsets, Busks and Clasps.

1882 PatGB5429. Katherine Jane Dance, Chadwick Road, County of Kent. Storing and Treating Grain for Bread Making in the manufacture of bread, biscuits and the like.

1882 PatGB5437. Margarethe Pohl and Otto Pohl, Liverpool, County of Lancaster. Improvements in the preservation of milk and in appliances and apparatus therefore.

1882 PatGB6119. Annie Hewett. Dish Covers or Appliances for Retaining the Heat of Joints of Meat.

1882 Married Women's Property Act – married women able to own property.

1883 Agnes Marshall opens her cookery school in Mortimer Street, London.

1884 PatGB5443. Sarah Marks (Hertha Ayrton), Spinster, Portsmouth, Hampshire. Improvements in Mathematical Dividing Instruments,

1884 Octavia Hill. Training scheme for women to manage mass housing projects for the Ecclesiastical Commissioners.

1884 Married Women's Property Act – women independent of their husbands.

1885 PatGB6528. Sarah Simsey, Housekeeper, Pembroke Street, Caledonian Road, Middx. Improvements in Washing Machines.

1885 PatUS322177. Sarah E. Goode, Owner of furniture store, Chicago, Illinois, USA. Cabinet bed.

1886 PatUS355139. Josephine Cochran, Chicago, Illinois, USA. Dish Washing Machine.

1886 PatGB4566. Margaret Louisa Corrie, Woking, Surrey. An improved Ankle Guard or Screen for Lady Cyclists.

1886 Agnes Marshall starts her weekly magazine, *The Table.*

1887 PatGB13125. Clara Louisa Wells, Spinster, c/o Messrs. W.J. Turner and Co., Naples, Italy. Improvements in Obtaining Fresh Water from Sea Water for Supplying Towns and for Other Purposes.

1887 PatGB15130. Annie McCleverty, Gentlewoman, Cromwell Road, South Kensington, Middx. An Improved Stand which may be used for a variety of purposes.

1887 PatGB16852. Marianne Catherine Stephenson, Spinster, Norfolk Street, Park Lane, London. Improvements in Music Stands and Reading Desks.

1888 PatUS386289. Miriam Benjamin, Schoolteacher, Washington DC, USA. Gong and Signal Chair for Hotels etc.

1889 PatGB2931. Elizabeth Barnston-Parnell, Metallurgist, St Johns Road, St John's, Kent. Improvements in the Treatment of ores for the recovery of metals and in apparatus therefor.

PatGB4354. Elizabeth Barnston-Parnell. Improvements in or relating to Calcining Furnaces.

1889 PatGB3138. Catherine Annesley Malcolmson, Gentlewoman, Norrysbury, East Barnet. An Improved Appliance for Holding or Suspending Closed Umbrellas, Sunshades or the like from the person of the wearer.

1889 PatGB12276. Mary Brown, Teacher, 4 Shaftesbury Place, London Road Station, Brighton, Sussex. Improvements in Coppers.

1890 PatGB1477. Amy Lange, 24 Mecklenburgh Street, Leicester. Appliances for Temporarily or Permanently Shortening a Dress.

1890 PatGB5917/13903. Susan Arabella Mackie, Gentlewoman, Chancery Lane, London. Improvements in and Connected with Boiling Apparatus for Washing and Laundry Purposes.

1890 PatGB9425. Mary Wallace McAllister, Dealer in Furniture, New City Road, Glasgow, Lanark, North Britain. A Combined Chiffonier, Cupboard or Chest of Drawers, and Bedstead and Bed.

1890 PatGB9622. Emma Pike, Gentlewoman, Trebovir Road, Earls Court, London. An Improved Bronchitis Kettle.

1890 PatGB9713. Sarah Helen Garwood, Married, Taswell Road, Southsea, Hants. Improvements in Darning and in Apparatus connected herewith.

1890 PatGB9885. Melinda May, Manchester, Iowa, USA. An Improved Hanger or Support for Pictures, Draperies and Wall Decorations.

1890 PatGB12884. Lady Mary Cecil Elizabeth Wilhelmine Gladstone, Great Barton Vicarage, Bury St Edmonds, Suffolk. Improvements in Travelling Trunks, Baskets and the Like.

1890 PatGB15850. Clara Louisa Wells, of no calling or profession, residing at Pompeii, Italy, to be addressed to the care of the English vice consul, Mr Frederick Turner, Monte di Dio, Naples. Improvements in Aerial Locomotion.

1890 PatGB17933. Eliza Turck, Artist, St George's Square, Primrose Hill, Middx. An Improved Mode of Ornamenting Fabrics and Garments.

1890 PatGB20497. Emma Grimes, Knapton Hall, North Walsham, Norfolk. An Improved Finger Protector for the use of Seamstresses.

1890 PatGB20674. Mabel Annie-Lovibond, Spinster, Newhall, Ardleigh, Essex. A New or Improved Removable Cap for Stick of Sealing Wax, and the said cap being applicable as a stamp or seal.

1890 PatUS426486. Margaret Wilcox, Chicago, Illinois, USA. Combined Clothes and Dishwasher.

1891 PatGB3853. Sarah Helen Garwood, Married woman, Taswell Road, Southsea. Spouts for use on Bottles, Jars, Vessels and the like.

1891	PatGB7623. Mary Ann Claxton, Percy Street, New Brighton, Chester. An Improved Combined Bed Table and Book Rest.
1891	PatGB9362. Isabella Grace Peckover, Vernon Place, Bloomsbury Square, London. A Sanitary Sink Basket.
1891	PatGB15567. Elizabeth Barnston-Parnell, Metallurgist. The Acacias, Sutton, Surrey. Furnaces.
1891	PatGB20079. Marian Eliza Phillips, Gentlewoman, Birchington Road, Kilburn, London. An Improved Ear Trumpet.
1891	**Factory and Workshops Act – safety and hygiene conditions, minimum working age raised to eleven.**
1892	PatGB1207. Sarah Symons, Milliner, Smith Street, Guernsey. Improvements in Stands for Hats.
1892	PatGB3271. Mary Sandes Landowner, Hungerford. The Island, Clonakilty, Co. Cork. Boot Fastening.
1892	PatGB7185. Adolf Tenison and Frances Sarah Tenison, Lady Uxbridge Road, London W. Appliance for Handling or Removing Oven Hot plates or Shelves.
1892	PatGB10190. Matilda Stewart Barron, Spinster, Swanmore, East Molesey. A New and Improved Combination Dress Stand and Fire Escape.
	PatGB14545. Matilda Stewart Barron. Improvements in Noiseless Coal Scuttles.
1892	PatGB19710. Mary Lucretia Kesteven, Lansdowne Terrace, Hampton Wick, Surrey. An Improved Carving Fork.
1892	PatUS473653. Sarah Boone, New Haven, Connecticut, USA. Ironing Board.
1893	PatUS494397.Variations US525241/32613/622092. Marie Tucek, New York City, USA. Improved Breast Supporter.
1893	**World Columbian Exhibition, Chicago. Women's Pavilion.**
1895	PatGB19900. Agnes Louisa Drury, Authoress, Rivers Street, Bath. Improvements in and Relating to Clothing for Infants.
1895	PatGB24208. Elizabeth Barnston-Parnell, Metallurgist, The Nest, Wallington, Surrey. Boilers and Kettles.
1895	Octavia Hill with Robert Hunter and Hardwicke Rawnsley. Foundation of the National Trust for Places of Historic Interest or Natural Beauty.
1896	PatGB13715. Clara Louisa Wells, Toulon, to be addressed to Vice Consul Anglais, M. Jouve, Toulon (le Var), France. Improvements in Aerial Routes and Connected with the Distillation, Storage and Supply of Water.
1897	PatGB12836. Clara Louisa Wells, of no calling or profession, Residing at Toulon (le Var) France. Centres Providing Means for Controlling and Utilizing Volcanic, Aqueous and Meteorological Forces.
1897	PatGB25289. Lady Mabel Marian Lindsey, Burcote House, Clifton-Hampden, Abingdon, Oxford. Improvements in or Connected with Vehicles for Carrying Bicycles.
1897	PatGB25405. Elsie Kate Josephson, Gentlewoman, Heath House, Manor Road, East Molesey, Middx. Improvements in or Relating to the Prevention of Fraud Concerning the Contents of Glass Bottles and the like.
1897	PatGB25522. Laura Biddle, Married woman, Princess Street, Leicester. Improvements in Toys.
1897	PatGB25534. Cicely Fulcher, Spinster, Kingdon Road, West Hampstead, Middx. Improved Method of and means for Hanging Skirts and the like.

1897 PatGB25631. Madame Roels, Widow, Laeken near Brussels. Improvements in Bedsteads, Stretchers and the like.

1897 PatGB25793. Madame Charles Delecroix, Gentlewoman, Forges d'Uzemain, Vosges, France. New or Improved Hamper or Receptacle for Bottles Containing Acids, Essences and the like.

1897 PatGB26054. Clarissa Eveline Jay, Gentlewoman, North Adelaide, South Australia. Improvements in Means for Affixing and Adjusting Umbrellas, Sunshades, and Similar Articles to Cycles.

1897 PatGB26102. Anna Krüger, Schoolmistress, Langestrasse, Baden-Baden, Germany. An Improved Process for the Production of Pliable and Elastic Bodies by Electrolysis.

1897 PatGB26212. Annie Bradley, Sick-Nurse, Hazelcroft, Killinghall, Ripley, Leeds. Improvements in Scissors.

1897 PatGB26416. Henrietta Pearce, Berners Street, London. Improvements in and Relating to Combined Ashtrays and Rests for Cigars, Cigarettes and the like.

1897 PatGB26428. Margaret Montgomery Grant, Artist in Oils, Queens Gate, Aberdeen. Improvements in Connection with Tobacco Pipes.

1897 PatGB26626. Laura Shepherd, Artist, Princes Square, Kennington Park Road, Kennington. Improvements in Drain Testing Apparatus.

1897 PatGB26639. Julia Dennis Richards, Mineral Water Manufacturer, St Thomas' Square, Newport, Isle of Wight. Improvement in Connection with Mineral Water Syphons.

1897 PatGB26747. Elizabeth Bartram and H. Gurney, Sheffield, Yorkshire. Improvements in or Relating to the Wheels of Cycles or other Vehicles.

1897 PatGB27042. Christina Elizabeth Rawbon, Longsight, Manchester. Improvements in Bottle Stoppers, Drawer Knobs and like articles.

1897 PatGB27142. Annie Elizabeth Bassett, Jubilee House, Holsworthy, Devon. Improvements in Means for Securing Buttons to Garments.

1897 PatGB27189. Florence Jarvis, Artist and Spinster, Orlando Road, Clapham Common, Surrey. An Improvement in Ladies' Dresses.

1898 PatGB15679. Clara Louisa Wells residing at Bordighera (Liguria) Italy. Secure Modes of Exploring the Cold and Hot Regions of the Earth by means of centres of elevation and depression, with reference to volcanic, aqueous, and meteorological forces, and of routes suspended with or without balloons.

1899 PatGB22790. Fanny Sophia Smith, Spinster, Travencore, Eastbourne, Sussex, and Astley Carrington Roberts, Surgeon, Badlesmere, Seaside Road, Eastbourne, Sussex. Improvements in Invalid Chairs.

1900 Edith Dawson, Enameller. *A Little Book of Enamels.*

1901 PatGB222. Margaret Ann Irving, Married Woman, Ontario Street, City of Victoria, British Columbia. Improvements in Garment Supporters.

1901 PatGB225. Agnes Williams Moller, Third Street, New York City, USA. Combined Sewing Machine Attachments and Sewing Implements.

1901 PatGB265. Annie Knowles, Married woman, Park Farm, Bewdley and John Field, Edgbaston. An Improved Dress or Garment Suspender.

1901 PatGB4066. Eliza Cutler, Wellington Road, Edgbaston, Birmingham. Improvements in Plate Racks and Drainers.

1901 PatGB4281. Lady Elizabeth Cumming. Army and Navy Mansions, Victoria Street, Westminster. An Improved Sanitary Towel.

1901 PatGB4453. Ellen Isabella Pennefather, Spinster, Victoria Square, Bristol. An Improved Method of Securing Cartridges in Cartridge belt.

1901	PatGB9900. May Rath, Married Woman, Seymour Terrace, Annerley, Surrey. Improvements Relating to Pneumatic and like Tyres for Vehicle Wheels.
1901	PatGB9934. Ethel Louisa Steele, Gentlewoman, The Cedars, Hartwell, nr Longton, Staffs. An Improved Fastener or Connection. Means for Coupling and Holding Together at the Waist a Lady's Blouse and Skirt or the like.
1901	PatGB10054. Annie Smith, Mercantile Clerk; Mary Hamilton Archer, Stock-keeper; George Benson, Sewing Machine Manufacturer, all of Cloth Fair, London. Hat Pin Protector.
1901	PatGB10119. Agnes Louisa Drury, Authoress and Widow. Late of Granville Place, Portman Square, London and now of Thames View, Broom Water, Teddington. Improvements in and relating to Clothing for Infants, Children and Adults.
1901	PatGB14580. Louisa Agnes Bella Carver, Spinster, Barnard's Hill, Muswell Hill, London. Improvements in Means for Fastening Fragile Articles in Boxes, especially millinery and the like.
1901	PatGB15408. Anna Regner, Married Woman, Pirmasens, Germany. Improved Low Shoe.
1901	**Death of Queen Victoria, accession of King Edward VII.**
1903	PatGB3571. Louisa Ousey, Belle Vue Villa, South Wimbledon, Surrey. Improvements in Hair-pins.
1903	PatGB3964. Sophie Vischer, W130 Street, New York City, USA. An Improved Body Garment or Support for the Bust and Abdomen and also for the Skirts and Hose.
1903	PatGB4253. Agnes Johnston Holmes, Regent Street, Ascot Vale, State of Victoria, Australia. Utensils for Removing Hot Dishes, Bowls and the Like from Ovens or for Holding such Articles while in a Heated Condition.
1903	PatGB4256. Sarah Lewis, Gentlewoman, Fern Villa, Johnstone, Carmarthen. Improvements in Devices for Preventing the Laces of Boots Shoes and the like from becoming unfastened.
1903	PatGB4312. Sophie Jackson, The Maisonette, Eastern Road, Brighton, Sussex. Improvements in and Relating to Trivets.
1903	PatGB4411. Nellie Eliza Johnson, Housekeeper and Homer Garrison, Merchant of Frankston, Anderson, Texas. Seams for Joining Fabrics, Metals, Papers and other Materials.
1903	PatGB4453. Anna Schaefer, Gentlewoman, Independence, Missouri, USA. Improved Seat, Carrier and Table Attachment for Baby Carriages and other Vehicles.
1903	PatGB4525. Mary Eliza Pritchard, Married lady, Boundary Road, St John's Wood, London NW. Improvements in Connection with Corset Busks.
1903	PatGB4577. Emma Moller, Gentlewoman, Guntherstrasse, Hamburg, Empire of Germany. Improvements in and Relating to Patterns for Cutting out Garments.
1903	PatGB4824. Marie Agnes Erembert Auzeric (née Armantier), Boulevard St Martin, Paris, France. Improvements in and Relating to Cooking Stoves.
1903	PatGB4882. Jane Ellen Stubley, Married woman. Carholme Road, Lincoln. An Improved Method for Supplying Better Light for Person when Reading in Railway Trains, Ships and Badly Lighted Rooms.
1903	PatGB4925. Katie Mary Moran, Seaford Lodge, Catford, London SE. A Scraper and Water Sprinkler for Cleaning Tram Lines or the Like.
1903	PatGB5156. Florence Ellen Gill, Elsinore Road, Forest Hill SE. An Improvement in or Relating to Bath Geysers and like Rapid Water Heating Apparatus.

1903 PatGB5311. Clara Hedwig Martini, Manufacturer, Bayerischestrasse, Leipzig, Germany. Improvements in and Connected with Fittings for Steam Heating, Cooking and like Apparatus.

1903 PatGB6711. Annie Chapman, Dressmaker, Park Lane, Southwold, Suffolk. Improvements in and Relating to Pin Cushions.

1903 PatGB6721. Janet Walker, Dressmaker, Adelaide Street, Brisbane, State of Queensland, Commonwealth of Australia. An Improved Dress Stand Figure or Dummy for Use in Dressmaking.

1903 PatGB6976. Amy Jane Booker, Married woman, Abbots Langley, Herts. Metal Mounted Tobacco Pipes.

1903 PatGB6997. Martha Kate Woodall, No occupation, Walden, Chislehurst, Kent. Improvements in Ladies' Skirts.

1903 PatGB7197. Maria Pressaud, née Meunier, Corset maker, Ebury Street, Eaton Square, London W. Improvements in Straight Corsets.

1903 PatGB7285. Magdalena Laue, Inventor, Merseburgstrasse Halle a/S Germany. Baby Napkins.

1903 PatGB7439. Amelia Margaret Elizabeth Summers, wife of Thomas Summers, Hooton Hall, Chester. An Improved Receptacle for Carrying Coats, Rugs and other Articles.

1903 PatGB7517. Edith Olive Whittlesey, Married woman, Oakley Avenue, City of St Paul, County of Romsey, Minnesota, USA. Improvements in Apparel Belts, Collars and the like.

1903 PatGB9141. Bertha Miller Windoes, Kalamazoo, Michigan, USA. Improvements in Globe Holders for use on gas burners.

1903 PatGB9379. Eleanor Clifford, Art Decorator, Western Road, Hove, Sussex. A Letter Card and Postcard Combined.

1903 PatGB9678. Edith Agnes Easthope Holdsworth, 24 St Ann's Villas, Holland Park, Kensington W. An Improvement in the Method of Fixing & Securing Covers of Sunshades & Umbrellas & Allowing of their Removal & Replacing with New at Will.

1903 PatGB9791. Frances Florinda Crawford, Gentlewoman, Dunnville, Ontario, Canada. Improvements in Apparatus for Preventing the Shrinking of Garments During Drying after Washing.

1903 PatGB10427. Emily Elizabeth Macaulay Mitchell, Spinster, Petit Tor Terrace, St Mary Church, Torquay, Devon. Improvements in and relating to the Walls of Tents, Huts and the like.

1903 PatGB12503. Augusta Florence Barnett, no Occupation, Kelvin Road, Highbury Park, London. Improvements in Tea Pot and Like Strainers.

1903 PatGB15117. Martha Helliwell, married woman, Spring Cottage, Utley, Keighley, York. Improvements in and Relating to Cowls for Chimney and Ventilating Shafts.

1903 PatGB15237. Clara Louisa Moore, curler, Clowes Street, Holbeck Leeds, Yorkshire. Improvement in Hair Curlers and the like.

1903 PatGB18298. Lady Alice Walkinshaw. Hartley Grange, Winchfield, Hants. Improved Device for Suspending Hats and other Articles from the Backs of Theatre Fauteuils and the like.

1903 PatGB19094. Baroness Margarethe Johanne Christianne Marie von Heyden, Westensee, Holstein, Germany. A New or Improved means for Preventing Coition in the Case of Bitches and other Female Animals.

1903 PatGB19038. Margarethe Blume, Lady Fachingen a/Lahnbahn, Germany. An Improved Machine for Labelling Bottles, Boxes or other Similar Receptacles.

1903 PatGB19157. Kate May Bargery, Certified Nurse, Mildmay Hospital, Austin Street, Bethnal Green, London. An Appliance for use with Feeding Bottles.

1903 PatGB19563. Elizabeth Gunning, Married Woman, Caldwell, Pikestone House, Downpatrick, Co. Down. Improvements in Socks and Stockings.

1903 PatGB19774. Anna Necker, Manufacturer, Adalberstrasse, Berlin, Germany and Dr Max Lassberg, Chemist, Steinwetzstrasse, Berlin. Process and Apparatus for Sewing Books.

1903 PatGB19892. Emily Caroline Faust of Kasson, Minnesota, USA. Improved Apparatus for use in Making Crochet Work Garments.

1903 PatGB21214. Isabella Scott, Hubbard Street, Jacksonville, Duval, Florida, USA. Improvements in Garment Hangers.

1903 PatGB21694. Eileen Isabella Dumville, No occupation, Red Hall Terrace, Vernon Avenue, Clontarf, Dublin. Improvements in and in Connection with Paint Brushes and the like.

1903 PatGB22263. Edith Ellen Louisa Marshall, Wife of Revd Robert Manning Marshall, Sloane Terrace Mansions, Chelsea, London. Improvements Relating to Saucepans and other Vessels or Receptacles having a handle and lid.

1903 PatGB22393. Harriet Barlow Heron, No calling, Devonshire Road, Bexhill-on-Sea, Sussex. A New or Improved Device for Facilitating the Cleaning of Windows and the like.

1903 PatGB22471. Clotilde Chotteau née Veldekeins, Gentlewoman, Rue de Naples, Paris, France and Emile Disse Cooper, Rue de Maise, Paris, France. Improvements in or Relating to Apparatus for the Manufacture and Storage of Wine Vinegar for Domestic Purposes.
PatGB23066. Clotilde Chotteau and Emile Disse Cooper. A New or Improved Ferment for the Manufacture of Wine Vinegar.

1903 PatGB22631, Florence Georgina Blake, Spinster, Pepys New Road, New Cross, Kent. Improvements in Connection with Outside Shop Blinds for Business and other Premises.

1903 PatGB22738. Kate Drusilla Stevenson, Married woman, Churchfield Road East, Acton, Middx. An Improvement in Collar Retainers.

1903 PatGB23063. Anna Emilie Louise Christens, Married woman, Norresogade, Copenhagen. Denmark. A Device for use in Closing Sacks.

1903 PatGB26221. Annie Charlotte Keightley, Decorator, Church Street, Kensington, Middx. Lewis Cockerell, 50 North End Road, West Kensington. Improvements in Bedsteads.

1903 GBReg.Design No. 423 888, BT51/118. Frederick Warne & Co., Bedford Street, Strand, London WC, Publisher. Design of Beatrix Potter's Peter Rabbit as a soft toy.

1903 PatUS743801. Mary Anderson, Birmingham, Alabama, USA. Window Cleaning Device.

1903 Nobel Prize for Physics Awarded to Marie Curie, Pierre Curie and Antoine Becquerel for the discovery of radium and its destruction of cells.

1904 PatGB1247. Matilda Soanes, no occupation, Lowestoft, Suffolk. Improvements in Mattresses.

1904 PatGB1567. Louisa Llewellin, Lady Holloway Road, London. Improvements in Gloves for Self Defence.

1904 PatGB1733. Ellinor Caroline Louisa Close, Married Woman, Eaton Square, County of London. Improvements in Fire-places or Stoves.

1904 PatGB2207. Caroline James, Wire Worker and Weaver trading as George F. James of Cleveland Wire Weaving Works, Wolverhampton,

	Staffordshire. Vincent Selvey, Foreman in the employment of the said Caroline James, of the same address. Woven Wire Articles such as Fire-guards, Fenders and the like, having the Strengthening Bands used therewith Fixed in an Improved Manner.
1904	PatGB2340. Amelia Jane Horne, Granville Park, Lewisham. Kate Santley, Lessee of the Royalty Theatre, Soho, London. New or Improved Method of Assisting Safe Exit from theatres and Other Buildings.
1904	PatGB2522. Mary Maria Collins, Costumier, Burnt Ash Road, Lee, London SE. Improvements relating to fastenings for the Busks of Corsets, or for Leggings, or Gaiters.
1904	PatGB2606. Marie Eugénie Eddy, Housewife, St Louis, Missouri. Improvements in Soap Receptacles.
1904	PatGB2639. Ada Byron Dugan, Spinster, Asylum Road, Londonderry, Ireland. An Improved Mattress for Invalids.
1904	PatGB3015. Alice Marion Leigh, Spinster. Vincent Stanley Leigh, Surveyor, Mare Street, Hackney, London. Improvements in or relating to Cases or Covers for Hats.
1904	PatGB3062. Nelly Grace Bacon, Journalist, Norfolk Street, London WC. Improvements in Window Curtain Supports and Hangars.
1904	PatGB3868. Emma Vashti Lake, Philadelphia, USA. Improvements in Garment Hooks.
1904	PatGB4574. May Tennant, Married woman, Bruton Street, London. Improvements in Air Filtering Ventilators.
1904	PatGB5414. Jane Gardner Hampson, No occupation, Southport, Lancashire. Improvements in and Relating to Coal Boxes.
1904	PatGB5924. Dora Buhlmann, Married woman, Kaiserslautern, Germany. Improvements in and connected with Heat Insulating Devices, Digesters and the like.
1904	PatGB6058. Maud Morton, Married woman, South Croydon, Surrey. Improvements in Pins used for Securing Hats or other Head Coverings in Position.
1904	PatGB6419. Theodora Wilson-Wilson, Spinster, Low Slack, Kendal, Cumbria. Improvements in Brooders or Artificial Mothers for Rearing Poultry and for other like purposes.
1904	PatGB6519. Annie Cullen Birkin, Married woman, Hanwell, Middlesex. Improvements in and Relating to Pneumatic Tyres.
1904	PatGB7189. Fanny Wassell Chilton, Married, Leytonstone, Essex. A Machine for Sifting Corn, Chaff and the like.
1904	PatGB7241. Margaret Clarke, Dress-cutting teacher, Sauciehall Street, Glasgow. Improved Dress and Costume Cutting Charts for Ladies and Children's Garments.
1904	PatGB8280. Mary Eliza Taylor, Laundry Manageress, Huddersfield, County of York. An Improved Guard or Hand Protector for Laundry Ironing Machines.
1904	PatGB8817. Miranda Malzac, Engineer, Process for Desulphurising and Oxidising Cadmium, Cobalt, Copper, Nickel, Silver, Zinc and like Ores by a Wet Method.
1904	PatGB9421. Lillian Caroline Sangster, Spinster, Ridgeway Road, Red Hill, Surrey. Improvements in Ruling Attachments for Measure of Length, being Two Instruments to Fix on measures, to Measure and Mark at same time.
1904	PatGB9607. Jane Hodgkinson, Married woman, Oxford Road, Manchester. Improvements in Abdominal Belts.

1904	PatGB10139. Lizzie Graessle, Housewife, San Jose, California, USA. Improvement in Combined Bed and Table.
1904	PatGB10265. Lavinia Laxton, Electrician, Brixton, London. An Improvement in and Relating to Spindles for Cycles Tricycles and other Velocipedes.
1904	PatGB15273. Susan Broadbent, Gentlewoman, Hinckley Road, Leicester. Improvements in or Relating to Corsets.
1904	PatGB16562. Mary Nice, Widow, Bury St Edmunds. Improvements in and Connected with Gas Stoves.
1904	PatGB16795. Mary Ann Charles, Costumier, Stapleton Road, Bristol. Improvements in Cooking Utensils for Vegetables and other Articles.
1904	PatGB17256. Mary Hill and John Hill, Jewellers and Gas Fitters, Trimdon Colliery, R.S.O. in the County of Durham. Improvements Relating to Acetylene Gas Generating Apparatus.
1904	PatGB17466. Annie Josephine Hebblethwaite, Spinster, Blenheim Terrace, Leeds. Improvements in and Relating to Drawers and the like for personal wear.
1904	PatGB17555. Jeannie MacNiel, Crosshill, Glasgow. Improvements in the Manufacture of Golf and other balls.
1904	PatGB17740. Mrs Bertha Hope, Hopedene, Wellingborough, Northants. An Improvement in Under-bodices.
1904	PatGB17873. Emily Gertrude White, Drapery Buyer, Fulham, London and Herbert Waterman Beall, Fancy Draper, High Barnet, Hertfordshire. Improvements in Combinations, and Similar Articles or Underwear for use with Stocking Suspenders.
1904	PatGB18028. Carla von Kropff, Baroness zu Putlitz, Riegnitz, Germany. Improved Incubator.
1904	PatGB18590. Florence Boggis, Professionally known as Mlle de Dio, Dancer, c/o Thomas Shaw & Co., 18 Adam Street, London WC. An Improved Method of and means for Displaying Stage Representations, Advertisements and for other purposes.
1904	PatGB18676. Florence Jules Heppe, Manufacturer, Chestnut Street, Philadelphia, Pennsylvania, USA. Improvements in Automatic Musical Instruments.
1904	PatGB18695. Elizabeth Margaret Johnson, Twine and Lace Merchant, Albemarle Road, Chorlton-cum-Hardy, Manchester, County of Lancaster. Improvements in Means or Apparatus for Supporting Cheeses or Balls of Twine and like material.
1904	PatGB19381. Adelaide Sophia Turner, née Claxton, Bath Road, Chiswick, Middlesex. A New or Improved Appliance for Curing 'Double Chin'.
1904	PatGB19531. Winifred Mundler, Married Woman, Haden Hill, Wolverhampton, Staffs. An Improved Device for use in Connection with Incandescent Gas Burners.
1904	PatGB19656. Johanna James Strain, Married Woman, Christchurch, Colony of New Zealand. Improvements in or Relating to Gas Stoves.
1904	PatGB20342. Emma Wyld, Married Woman, Shepperton, Surrey. Combined Automatic Beer Saving Apparatus and Filter.
1904	PatGB20442. Julia Batcheller, Gentlewoman, Maidstone, Kent. An Improved Polish for Metals.
1904	PatGB20718. Annie Flora Catherine England, Married Woman, Maida Vale, Middx. An Improved Spoon.
1904	PatGB20756. Flora Marie Duchesse de Touraine & Douglas née Hamilton, St Germain des pres Paris, France. A Portable Wardrobe.

1904 PatGB20982. Ellen Rogers Pickthall, wife of Mr A. Pickthall, umbrella maker. Hanley, Staffs. A Flycatcher.

1904 PatGB21267. Sarah Ellen Bazeley, Married Woman, Cafe Buildings, Abington Square, Northampton. Improved Means for Indicating the Deflation of Pneumatic Tyres, Produced by Puncture or otherwise.

1904 PatGB21351. Elizabeth Keswick, Gentlewoman, Wedholme, Skipton Road, Ilkley, York. Improvements in or relating to Apparatus for use in connection with Enemas, or Irrigators, and the like Injectors.

1904 PatGB21649. Emily Cohn, Bayswater, London. Improvements in Collars for Personal Wear.

1904 PatGB22349. Clara Cunningham, Washington DC, USA. Improvements in Guides for Fur and Carpet Sewing Machines.

1904 PatGB22530. Katherine Tojetti, Manufacturer, 60 W75 Street, New York City, USA. Improvements in and relating to Spring Fastening devices.

1904 PatGB25849. Jane Horlick, Married Lady. Rugby, Warwickshire. Improvements in Suspenders for Coats, Blouses and the like.

1904 PatGB25906. Caroline Lucy Claypole, trading as the Armstrong Manufacturing Company, 38 Wilton Street, Finsbury, London EC. Improvements in and relating to Self-closing or Retractile Seats or Chairs.

1904 PatGB26012. Veuve Anna Marie Leroux, Rennes, Republic of France. Modifying Apparatus for the Mouth.

1904 PatGB26024. Edith Maria Mountford, wife of James Mountford, Solihull, Warwickshire. Improvements in Stiffeners for use in connection with certain articles of Wearing Apparel.

1904 PatGB27650. Bertha Chmielewsky or Sherwinter, Buccleuch Place, Edinburgh. Douche Back-rest or Bed.

1905 PatGB3052. Eliza Mitchell, Married woman, 'Verona', Worthing. Improvements in and connected with Umbrellas, Sunshades, Walking Sticks and the like.

1905 PatGB3226. Anna Lawson, Married, Northumberland Avenue, Dublin. Improvements in Pudding Bowls.

1905 PatGB3492. Lady Judith Fellheomer. Wooster Street, City of New York (Borough of Manhattan), USA. Improvements in Corset Covers.

1905 PatGB3596. Johanna Khabe née Schonfeldt, Berlin, Germany. Improvements in and relating to Tailors' Irons and the Like.

1905 PatGB6459. Beatrice Kent, Trained nurse, Nurses' Hostel, Francis Street, London. Suspensory Device for Retaining Surgical Bandages and the like, in Position on the Patient.

1905 PatGB6485. Justine Lewis-Thiessen, Married lady, St Gilles, Near Brussels, Belgium. Improved Device for Vehicle-wheels for Preventing Slide-slipping or skidding thereof.

1905 PatGB6646. Florence Mary Lilley, Ealing, Middx. Improvements in or relating to Spreading Devices for Maintaining Neck Bands, Dress Collars, Cuffs or other Pieces of Flexible Material in an Extended State.

1905 PatGB8336. Mabel Ahronsberg, Spinster, Portland Road, Birmingham. Improvements in Bows or Ties for Personal Wear.

1905 PatGB8607. Marian Borsa, Married Woman, Albert Park, State of Victoria, Commonwealth of Australia. An Improvement in or Connected with Stockings.

1905 PatGB9310. Sarah Annie Stewart, Ormskirk, Lancashire. An Improved Collar, Cuff or Belt Support for Ladies' Use.

1905 PatGB9424. Lillian Carpenter, Married Woman and Engineer, The Mechanic's Arms, Deptford, Kent. John James Holmes, St John's Road, Deptford. An Improved Tap or Cock.

1905	PatGB9521. Maria Vermeulen née Van Der Paelt, Brussels, Kingdom of Belgium. Abdominal Belt for Enceinte or Lying-in Women and Abdominal Sufferers combined with Bust-supporter and Waist Band.
1905	PatGB9698. Madeleine Kness, Tunis, Africa. A Process for the Transformation of Alfa and other similar Plants into Paper Pulp or textile Fibres.
1905	PatGB10013. Mary Martha Dunstone and Emma Ann Bartlett, Brixham, Devon. A New or Improved Sanitary Metal Commode.
1905	PatGB10122. Emma Thurgood, Bayswater, London and Leonard Haselwood West, Kensington, London. An Improved Grater for Nutmegs and other Comestibles.
1905	PatGB10192. Susannah Jane Parr, Married Woman, Wellingborough Northants. Improvements Relating to Hat Pins for Ladies' use.
1905	PatGB10616. Josephine Marie-Louise Fleming née Imbert, Château des Ormes sur Voulzie, France. Improvements in means for Generating and Using Hydrocarbon Vapours for Heating and Lighting Purposes.
1905	PatGB10866. Pauline Grayson, Artist, Victoria Street London. Improvements in Connection with Artificial Fuel.
1905	PatGB11221. Amelia Hauser Sinsheimer, New York, USA. Improvements in Garment Hangers.
1905	PatGB11590. Angelica Maria Anna, Countess Sponneck-Mayer, Kensington, London. Improvements in Elastic Fluid Engines of the Revolving Cylinder Type.
1905	PatGB11904. Emma Smith, No occupation, Weston-super-Mare, Somerset. Improvements in Writing Desks for the Blind and Persons having Impaired or Defective Vision.
1905	PatGB12025. Edith Emily Whittaker, Medical and Maternity Nurse, registered CMB London, Liscard, Cheshire. An Improved Nipple Shield.
1905	PatGB12148. Annie Louisa Proctor, Gentlewoman, Manchester. An Improved Means for Carrying Umbrellas, Parasols, Walking Sticks or the Like.
1905	PatGB12340. Louise Fletcher, Gentlewoman, no occupation, Euston Square, London. Ladies' Veil retainer.
1905	PatGB12410. Clara Allertz née Madel, Married Woman, Düsseldorf, Germany. A Composition of Matter for Cleaning and Polishing Metal Plates.
1905	PatGB12428. Sarah Ann Norledge, Matron of Nursing Home, Hawkwood Road, Boscombe, Hampshire. An Improved Extension Apparatus for Surgical Purposes.
1905	PatGB12466. Mary Cecilia Collier, Washington DC, USA. Improvements in Folding and other Cribs.
1905	PatGB12272. Annie Florence Wilson Bowen, Housewife, San Francisco, California, USA. Rubber Holding Attachment for Pencils.
1905	Nobel Peace Prize awarded to Baroness Bertha Sophie Felicita von Suttner.
1905	**Patent Act**
1906	PatGB647. Nellie Digney, Dressmaker, Bridgeport, Connecticut, USA. Improvements in Combined Abdominal Supports and Hose Supporters.
1906	PatGB762. Florence Elizabeth Herndon, Dallas, Texas, USA. Improvements in Hats.
1906	PatGB1548. Anna Hetterman, Artist, Schoneberg, near Berlin, Germany. Improvements in and relating to Chairs for Public Buildings.
1906	PatGB6386. Emily Wynne-Jones, Certified Mistress. Belgrave Road, Ilford, Essex. Improved Kinder-garten Tree.

1906	PatGB7989. Isabelle Johnson, Gentlewoman, Cokedale, Montana, USA. Improvements in Combined Body Braces or Stays and Garment Supporters.
1906	PatGB8065. Henrika Båhr, Teacher, Helsingfors, Finland. An Improved Fountain Penholder.
1906	PatGB8118. Augusta Pattersson, Trained Certified Teacher, Fillebrook Road, Leytonstone. An Apparatus for the Thorough Teaching of Numbers from one to twelve inclusive.
1906	PatGB11186. Elizabeth Reid, Lady Hope, Connaught Place, Hyde Park. Improvements pertaining to hats securing them to the head.
1906	PatGB20363. Laura Edwards and Edward Edwards, Commission Agent, Pontypridd. A New Appliance for use in connection with Mantelpieces and like Supports as a Bath Tub Screen, Fire Screen, Draught Screen and Clothes Horse for other purposes.
1906	PatGB28639. Georgiana Scott and Anna Louisa Scott, Mona Cottage, Ossett, York. A New or Improved Device or Means for Attaching or Securing Cords to Window Blinds, Shades or the like.
1906	PatGB29186. Florence Zambra, Church End, Finchley, Middlesex. A Toy garden.
1906	Edith Dawson, Enameller. *A Little Book of Enamels.*
1907	PatGB757. Sophia Lucy Caroline Barnacle, Married Woman, Hermitage School, Grimsaugh, Preston. Improvements in Amusement Apparatus.
1907	PatGB1312. Isabella Parker, Spinster, Annfield Plain, County of Durham. Improvements to Devices for the Inserting into the Human Ear to thereby artificially increase the Hearing of Persons whose Natural Ear Drums are Defective.
1907	PatGB2053. Bertha Trautmann née Sprie, Zollnerstrasse, Dresden, German Empire. Improvements in Electrical Apparatus for Automatically Indicating the Positions of Trains on Railways.
1907	PatGB2785. Mrs Louisa Davis Brown, Artist, Kelvingrove, Upper Sydenham, London SE. Improvements in Funeral Carriages and Hearses.
1907	PatGB3099. Alice Maude Nelson Phipps, Married Woman, Cliftonville, Northampton. An Improvement in Ladies' Hatpins.
1907	PatGB3126. Amy Winifred Zwicker, Married Woman, Catford, County of Kent. Improvements in Personal Sanitary Conveniences for Invalids and others.
1907	PatGB3346. Elisabeth Beckmann, born Shulte, Baerwaldstrasse, Berlin, German Empire. Improvements in Cylinder Washing machines.
1907	PatGB5166. Florence Crosby Hurd, Married Woman of no occupation, Southwood Avenue, Highgate, Middx. Improvements in Appliances for Inducing Correct Breathing through the Nose instead of the Open Mouth; for Flattening Prominent Ears and Keeping Surgical Bandages and the like in place on the Head.
1907	PatGB21322. Catherine Dowie and Charlotte Easson, Newlands, Glasgow. Improvements in Collapsible Tables.
1907	PatGB21841. Barbara Dale Howarth, Compton Road, Wolverhampton, County of Stafford. Improvements in Ear-rings.
1907	PatGB23296. Mabel Mason Van Vechten, Registered Nurse, W45 Street, Manhattan, County and State of New York. Improvements in and Relating to Side Saddle.
1907	PatGB23574. Kate Morgan trading as 'Kate Wolstenholme', Wells Street, Oxford Street, London. Corset and Surgical Belt Maker. Improvements in Apparatus for Housing and Applying Step Carpets or the like in Motor and other Vehicles, and for like use.

1907	PatGB23576. Alice Lang, Wellington Road, Edgbaston, Birmingham. Occasional Sanitary Shield.
1907	PatGB23950. Grace Flora Woods, Married Woman, Ackers Street, Chorlton-on-Medlock, Manchester. Improvements in or applicable to Electroliers and in Ceiling Roses therefore.
1907	PatGB24909. Anna Louisa Raw, Rodney Street, Liverpool. A Portable or Folding Cot.
1908	PatGB1829. Grace Elizabeth Markell, Gentlewoman, of no permanent address, residing temporarily at Hotel Longfellow, Boston, Massachusetts, USA. Improvements in Girdles, for Personal Wear.
1908	PatGB1992. Fanny Louisa Flower, Married woman, Sandy Grove, Pendleton, County of Lancaster. A New or Improved Apparatus for Enabling Glasses, Cups, Plates and the like to be washed without wetting the hands.
1908	PatGB2950. Ella McAllister Bennett, Gentlewoman, Sulphur, State of Oklahoma, USA. Improvements in and Relating to Kitchen Cabinets.
1908	PatGB3570. Ursula Sofie Baroness Bahlerup, Charlottenlund, County of Copenhagen, Denmark. A New or Improved Mat or Mattress.
1908	PatGB4016. Mary Shippobottam, Entertainment Contractor, Orlando Street, Bolton, County of Lancaster. Improvements in Roundabouts.
1908	PatGB6965. Emma Susan Barwell, Married, Edgbaston, Birmingham. An Improved Abdominal Belt or Support.
1908	PatGB6937. Mrs Bertha Ortell, Gentlewoman, East 116 Street, New York City, USA. Combined Pocket and Garment Supporter.
1908	PatGB8839. Ada Anna Ashmall-Salt, Spinster, Marlborough College, Buxton, County of Derby. An Improved Pudding Mould.
1908	PatGB9649. Alexandra Richter, Gentlewoman, Hospitalstrasse, Leipzig, Germany. Improved Sterilising Apparatus.
1908	PatGB9837. Mary Jackson, Married Woman, Aintree, Liverpool. A Flat-folding Chair.
1908	PatGB10605. Emily Curtis, Gentlewoman, Glemsford, Suffolk. Improvements in Domestic Aprons.
1908	PatGB10618/10619. Muriel Binney, Married Woman, Sydney, Australia. Improvements in Cots for Children/A Portable Safety Playground for Infants.
1908	PatGB12467. Emily Hettie Canham, Married Woman, Highbury, London. Improvements in Lenses for Motor Car Lamps and Vehicle Lamps.
1908	PatGB13084. Ellenor West, Guildford, Surrey. Ointment.
1908	PatGB14679. Anna Lohr and Karl Romstaedt, Berlin, Germany. Trading as Lohr and Romstaedt. Improvements in Air Vessels for water Supply Pipes.
1908	PatGB15653. Florence Barker, Hyde Park, Leeds, County of York. An Improved Round Game.
1908	PatGB15747. Anna Holzweissig, Tradeswoman, Elberfeld, Germany. Improvements in or Relating to Devices for Protecting Confectionery during transit.
1908	PatGB16035. Isabella Mary O'Connor, Nurse, Lower Leeson Street, Dublin, Ireland. Improvements in Hospital Urinal Vessels and the like.
1908	Utility Model (Berlin) 347 896. Melitta Bentz, Dresden, Germany. Filtration-paper-based coffee filter with rounded and recessed bottom perforated by slanting flow through holes.
1908	Helena Rubinstein opens beauty salon in London.

1908 Elizabeth Arden salon opened by Florence Graham, 5th Avenue, New York City, USA.

1909 PatGB6930. Maud Walkey, East Sheen, London SW. An Apparatus for the Preservation of Milk by Scalding or Pasteurising.

1909 PatGB9083. Emma Stevens, Married Woman, Casson Gate, Rochdale. Improvements in Curtain Holders.

1909 PatGB9944. Jane Hampson, Promenade, Southport. Novel Apparatus for use in Learning or Practising Roller Skating or 'Rinking'.

1909 PatGB10313. Maude Coe, Married Woman, Oxford Street, London W. Improvements in Waistcoats for Men's Wear.

1909 PatGB10405. Kathleen Anglim, Grocer, Dulwich Hill, Sydney, Australia. Improvements in Lids or Closures for Earthenware or Glass Containers.

1909 PatGB10531. Jane Stewart Spiers, Maria Carstairs Spiers and Mary Stewart Jackson née Spiers, steel plane manufacturers, trading as Stewart Spiers, Old Bridge Street, Ayr. William McNaught, Foreman. Improvements in Planes for Wood-working Purposes.

1909 PatGB11582. Daisy Hastings, Knowle, Bristol. Improved Device for Making Woollen Boas or Necklets.

1909 PatGB11643. Mary Clark, Boston, Massachusetts, USA. Scouring or Cleaning device for Cutlery.

1909 PatGB11738. Valeska Martin, Breslau, Germany. An Improved Pressure Relieving Device for the Toe Joint.

1909 PatGB12492. Kate Jenkins, Sydney, Australia. An Improved Floating Buoy or Receptacle for Carrying Valuables, Food, or other Necessaries in Case of Shipwreck.

1909 PatGB21036. Hedwig von Rzewuska née von Konopacka, of Brixen, South Tyrol. An Improved Device for Preserving the Coiffure.

1909 PatGB27449. Hermance Edan, Spinster, Paris. A New or Improved Game.

1910 PatGB1218. Annie Keightley, Decorator, 17 Church Street, Kensington. Lewis Cockerell, Decorator, 46 North End Road, West Kensington. Improvements in and Relating to Extendible Bedsteads.

1910 PatGB1230. Bertha Stahlecker, Married Woman, 87 Koenig Strasse, Cannstadt, Wurtemberg, Germany. Improved Door Fastening and Signal.

1910 PatGB1478. Eliza Amphlett, Clarendon Crescent, Leamington, County of Warwick. Improvements in and relating to Invalid or Bed Tables also suitable for similar Articles of Furniture.

1910 PatGB2549. Emily Clark, Prinstead, Harold Road, Margate, Kent. Improvements in or relating to Holdalls or Bags.

1910 PatGB2668. Eliza Williams, Married Woman, Abbotts Brooks, Bourne End, Buckinghamshire. A New Material for Making Under Clothing and other such Light Articles of Wearing Apparel.

1910 PatGB2675. Clara Wills, Scarisbrick Street, Southport. An Improvement in Ovens, Stoves and Ranges Heated by Gas.

1910 PatGB2718. Bertha Wells, Gentlewoman, Waterloo Street, Hove, Brighton. An Improved Mustard Pot.

1910 PatGB2868. Lucy Appleby, Married Woman, Penn, Wolverhampton, Staffordshire. Improvements in and Relating to Cable and like Gripping Mechanism.

1910 PatGB5908. Sheila O'Neill, Motor Car Driver, c/o Salters' Hall, Cannon Street, City of London. Improvements in Flying Machines.

1910 PatGB14406. Luise Kopmann (née Kreiensen), 4 Wolfrangerstrasse, Cassel, Germany. Improvements in or Connected with Cooking Saucepans.

1910 Accession of King George V.

1911	PatGB4475. Bertha Paatsch, Grüssau, Germany. Improved Machine for Cutting Bread into Slices and Buttering same.
1911	PatGB18118. Alwine von Otterstedt née Mellinghaus, Gentlewoman, Rhondorferstrasse, Rhondorf, near Hannef on the Rhine, Germany. Improvements in Ladies Hat Fasteners.
1911	PatGB18760. Edith Waltham, Married Woman. John Waltham, accountant. Tankerville Road, Streatham, London. Improvements in Stoppers for Bottles.
1911	PatGB19293. Mary Sutton, Married Woman, Bowen Terrace, Brisbane, Queensland, Commonwealth of Australia. Improvements in Pneumatic Cushion Springs for Wheeled Vehicles.
1911	PatGB19772. Eduard Tüchler, Mechanic, Marie Tüchler née Kretchmer, 63 Klosterneuburgstrasse, Vienna, Austria, Johann Boi, Hairdresser, Bernhard Kreindler, Merchant. Device for Securing Ladies' Hats in position on the head.
1911	PatGB20131. Martha Marx, Breslau-Kleinburg, Germany. Improvements in Receptacles for use in Baking Cakes.
1911	PatGB20307. Mary Witton, Spinster, Goldney Road, Maida Vale, London. Improvements Relating to Corsets.
1911	PatGB20548. Clara Wilson, Married Woman. George Wilson, Builder, Preston Street, Brighton. Improved Mud Guard for Motor Buses and other Vehicles.
1911	PatGB20601. Maria Werbatus, Gentlewoman, Bad Elster, Germany. Improved Method of Making Wax Flowers.
1911	PatGB22232. Hedwig Godefroid, Friederichstrasse, Berlin. An Electrically Driven Receipting and Cash Registering Till.
1911	Nobel Prize for Chemistry. Marie Curie. Application of radium in treatment of cancer. Radium Institute.
1911	Käthe Kruse, Germany. Manufactures dolls.
1912	PatGB5021. Lydia Martell, Married Woman. Edward Payne, Gentleman. Seattle, State of Washington USA. Improvements in Bust Supporters.
1912	PatGB5078. Mary Lumley, Professional Nurse, Gloucester Street, Warwick Square, London. Improvements in Abdominal Belts.
1912	PatGB5101. Alice Rawlins, No profession, Bray, Berkshire. An Improved Portable Wardrobe.
1912	PatGB 5585. Kate Braidwood, Matron, Infectious Hospital, Colchester. Improved Clinical Apparatus for Control of Temperature.
1912	PatGB5955. Marie Seligman, Spinster, Adelaide Road, London NW. Completely Collapsible Table.
1912	PatGB6706. Maria Montessori, Doctor, Via Principessa Clotilde, Rome, Italy. Improvements in or Relating to Apparatus for use in Teaching Children.
1912	PatGB7007. Noeme Cinotti, Spinster, Rue d'Assas, Seine and Oise, France. Noeile Delmazures née Cinotti. Yvonne Cinotti, Spinster. Improvements in, or connected with the Manufacture and Production of Artificial Flowers, Foliage and Fruits.
1912	PatGB25380. Elizabeth Countess de Gasquet-James, née Tibbits Pratt, Dresden, Germany. Improvements in or Relating to Foodstuffs for Military and like use.
1912	USD43680. Rose O'Neill Wilson, Taney, Missouri, USA. Kewpie® doll (Registered design).
1913	PatGB2873. Clara Kunze née Lange, Gentlewoman, 76 Kaiserstrasse, Breslau, Germany. Improved Device for Locking Wearing Apparel on Suspension Hooks.

1913 PatGB8171. Else Schmidt née Walter, Married Woman, Berlin-Tegel, Germany. Improved Machine for Cutting Loaves of Bread.

1913 PatGB13617. Ida Matzen née Timm, 35 Hermannstrasse, Alt-Rahlstedt, Schleswig-Holstein, Germany. Improved Protector for Ladies' Hats and the like.

1913 PatGB17890. Maria Montessori, Doctor, Via Principessa Clotilde, Rome, Italy. Improvements in or Relating to Apparatus for Teaching Children Arithmetic.

1913 PatGB18432. Jessie Deans, Spinster, Upper Norwood, London SE. Improvements in or Relating to Night stools, Commodes and the like.

1913 PatGB19472. Elizabeth Bolton, Gentlewoman, Demesne Road, Manchester and Tom Gray, Engineer, Deansgate, Manchester. Improvements in Steam Superheaters.

1913 PatGB20950. Margaret Kay, Elementary School Teacher, Kelmscott Road, New Wandsworth, London, SW. Means for Teaching Arithmetic.

1913 PatGB26136. Grace Maitland, Married woman, Bentley, Hampshire. A Convertible Trunk.

1913 PatGB26237. Eva Balfour, Married Woman, Bromley, Kent. Improvements in or Connected with Spikes.

1913 PatGB26274. Grace Nott, Married Woman, Collapsible Hammock Chair.

1913 PatGB26989. Florence Santer, Trading as C.J. Cuthbertson, Cheapside, London EC. Improvements in Automatic Apparatus for Causing the Extension and Contraction of Opera Hats and the like for Advertising Purposes.

1913 PatGB27137. Anne Fitzgerald, Married Woman. Edward Fitzgerald, Sussex Mansions, Sussex Place, London SW. Improved Apparatus for Recording Number of Telephone Calls.

1914 PatGB11589. Lady Sophia Hall and Esther Woods, Directors of Mrs Oliver Ltd., Dressmakers, 39 Old Bond Street, London. Improvements in and Relating to Sleeping Blankets.

1914 PatGB14481. Maria Montessori, Doctor, Via Principessa Clotilde, Rome, Italy. Improvements in or Relating to Apparatus for Teaching Children Geometry.

1914 PatGB20707. Pauline Woolf, Secretary, Anchor Works, Windmill Street, Birmingham. Improvements in the Construction of Teats or Soothers.

1914 Dora Lunn, Ceramicist, Ravenscourt Pottery, London.

1914 **Outbreak of First World War, August.**

Notes

Chapter One

1. Patricia Phillips, *The Scientific Lady*, Weidenfeld & Nicolson, 1990.
2. Asa Briggs, *Victorian Things*, Penguin, 1988.
3. The UK Patent Office, 2002.
4. Stephen van Dulken, *Inventing the 19th Century*, British Library, 2001.
5. Stephen van Dulken, *British Patents of Invention 1617–1977*, British Library, 1999.

Chapter Two

1. Pat1637/GB104.
2. 1707 Act of Union of Great Britain.
3. Pat1678/GB204.
4. Pat1715/GB 401.
5. 'A new Invencon found out by Sybilla his wife, for Cleaning and Curing the Indian Corn Growing in the severall Colonies in America.'
6. Pat1716/GB403.
7. Autumn Stanley, *Mothers and Daughters of Invention: Notes for a Revised History of Technology*, Scarecrow Press, 1993.
8. Linda Colley, *Britons Forging the Nation*, Pimlico, 1992.
9. Claire Tomalin, *The Life and Death of Mary Wollstonecraft*, Weidenfeld & Nicolson, 1974.
10. Amanda Foreman, *Georgiana Duchess of Devonshire*, Flamingo, 1999.
11. Alison Kelly, *Mrs Coade's Stone*, The Self Publishing Group and the Georgian Group, 1990.
12. Foreman, *Georgiana*.
13. Pat1800/GB2457.
14. Pat1808/GB3129.

Chapter Three

1. Sarah Levitt, *Victorians Unbuttoned: Registered Design for Clothing, Their Makers and Wearers 1839–1900*, Harper Collins, 1986.
2. Pat1716/GB403.
3. Mary Kies of Connecticut. Patent destroyed – Weaving Straw into Silk or Thread.
4. Pat1815/GB3930.
5. Stephen Bunker, *Strawopolis, Luton Transformed 1840–1876*, Bedfordshire Historical Record Society and Luton Museum 1999.
6. Now Royal Society of Arts; the Society had always welcomed women members, although they were not allowed to sit on committees.
7. Society of Arts Transactions, vol. 42.
8. Derek Hudson and Kenneth Luckhurst, *The Royal Society of Arts 1754–1954*, John Murray, 1954.
9. Pat1828/GB5630.
10. Report of Chief Inspector of Factories and Workshops, 1891.
11. Pat1863/GB2804.
12. Pat1863/GB1794.
13. 'Brevet no. 65090, en date du 10 novembre 1864, A Mme veuve Durst et M.Munt, pour dessins imprimés en couleur sur chapeaux et autres articles de feutre.' Luton museum has a hat block similar to the one described by Mme Durst-Wilde with its elegant and wavy brim. This is a hat for a sophisticated lady.

14. Leigh Summer, 'Working Class Women and Corsetry', *Costume* 36 (2002).
15. Pat1838/GB7640 and Pat1852/GB1064.
16. *Health and Beauty*, 1864.
17. Arthur Benson and Viscount Esher (eds), *The Letters of Queen Victoria*, vols I, II, III, John Murray, 1908.
18. Pat1855/GB726.
19. Pat1861/GB3216.
20. Pat1905/GB9521.
21. C. Willet Cunnington, *English Women's Clothing in the Nineteenth Century*, Dover, 1990.
22. Pat1861/GB2048.
23. Pat1863/GB2168.
24. Autumn Stanley, *Mothers and Daughters*.
25. 'Working Class Women and Corsetry', *Costume*.
26. Pat1903/GB3964.
27. Pat1893/US494397.
28. Stephen van Dulken, *Inventing the 19th Century*, British Library, 2001.
29. Pat1912/GB5021.
30. Pat1904/GB17873.
31. Pat1905/GB8607.
32. Pat1908/GB1829.
33. Pat1904/GB15273.
34. Pat1908/GB693.
35. Pat1861/GB3059.
36. Joy Emery, University of Rhode Island, Pattern Research Archive.
37. Pat1904/GB7241.
38. Pat1903/GB6721.
39. Pat1864/GB2301.
40. Pat1870/GB2662.
41. Pat1865/GB3298.
42. Pat1868/GB26.
43. Patent Abridgement Class 134, Umbrellas, British Library.
44. Pat1839/GB8086.
45. Pat1867/GB1951.
46. Pat1897/GB27189.
47. Pat1903/GB6997.
48. Pat1901/GB10054.
49. Pat1907/GB3099.
50. Pat1906/GB11186.
51. Pat1911/GB18118.
52. Pat1911/GB19772.
53. Pat1909/GB21036.
54. Pat1913/GB13617.
55. Pat1905/GB12148.
56. Pat1856/GB2330.
57. Pat1866/GB1446.
58. Pat1904/GB19381.
59. Pat1904/GB26012.
60. Pat1901/GB4281.
61. Pat1907/GB23576.
62. Pat 1904/GB17466.

Chapter Four

1. Chris Cook, *Britain in the Nineteenth Century*, Pearson Education, 1999.
2. John S. Deith, 'Mrs Agnes B. Marshall', in *Cooks and Other People*, ed. H. Walker, Prospect Books, 1995.
3. Hardyment, *From Mangle to Microwave*, Polity Press, 1988.
4. Pat1809/GB3221.
5. Hardyment, *From Mangle to Microwave*.
6. Pat1854/GB146.
7. Pat1859/GB763.
8. Pat 1903/GB9141.
9. Pat1905/GB10616.
10. Pat1803/GB2702/2703.
11. Pat1880/GB1918.
12. Pat1892/GN14545.
13. Pat1903/GB15117.
14. Hardyment, *From Mangle to Microwave*.
15. Pat1903/GB5156.
16. Pat1906/GB20363.
17. Pat1862/GB2712.
18. Pat1878/GB2213.
19. Pat1913/GB18432.
20. Pat1861/GB958.
21. Pat1857/GB1410.
22. Pat1855/GB2837.
23. Pat1859/GB1778.
24. Pat1889/GB12276.
25. PatUS101321.
26. Pat1907/GN3346.
27. Pat1903/GB9791.
28. Pat1886/US355139.
29. Van Dulken, *Inventing the 19th Century*.
30. Pat1886/ US426486.
31. Pat1908/GB1992.
32. Pat1891/GB9362.
33. Hardyment, *From Mangle to Microwave*.
34. Pat1869/GB778.

35. Pat1903/GB4824.
36. Pat1892/GB7185.
37. Pat1863/GB1948.
38. Pat1812/GB3549.
39. Berlin 1908/ Gebrauchsmuster341.347896i1, 343556, 347895–96.
40. Melitta Unternehmensgruppe, Minden, Germany.
41. Pat1637/GB104.
42. Pat1859/GB2573.
43. Pat 1873/US139580/140508.
44. Inventors' Museum.
45. Pat1909/GB6930.
46. Pat1904/GB6419.
47. Pat1864/GB1880.
48. Pat1897/GB25405.
49. Pat1897/GB26639.
50. John Deith, 'Mrs Agnes B. Marshall' in *Cooks and Other People*.
51. Pat1904/GB5924.
52. Hardyment, *From Mangle to Microwave*.
53. Pat1908/GB2950.
54. Pat1862/GB1760.
55. Pat1905/GB10122.
56. Pat1905/GB3226.
57. Pat1908/GB8839.
58. Pat1910/GB2718.
59. Pat1910/GB14406.
60. Pat1903/GB22263.
61. Pat1903/GB12503.
62. Pat1913/GB8171.
63. Pat1913/GB8171.
64. Nikolaus Pevsner, *High Victorian Design*, Architectural Press, 1951.
65. Pat1839/GB8234.
66. Pat1860/GB3116.
67. Pat1848/GB12075.
68. Pat1831/GB6186.
69. Pat1854/GB2202 and 1855/GB790; van Dulken, *Inventing the 19th Century*.
70. Pat1907/GB21322.
71. Pat1885/US322177.
72. Pat1890/GB9425.
73. Pat1891/GB7623.
74. Pat1903/GB26221.
75. Society of Arts Mechanics Award to Mr J. Goode.
76. Pat1904/GB3062.
77. Pat1909/GB9083.
78. Pat1904/GB4574.
79. Pat1904/GB4574.
80. Pat1903/GB22393.
81. Pat1892/GB10190.
82. Anthea Callen, *The Angel in the Studio*, Astragal Books, London, 1979.
83. Callen, *Angel in the Studio*.
84. Pat1903/GB19094.

Chapter Five

1. Katie Hickman, *Daughters of Britannia*, Flamingo, 2000.
2. Girls' Public Day School Trust, *A Centenary Review*, GPDST, 1972.
3. Pat1801/GB2485.
4. Museum of Childhood. Edinburgh, November 1986.
5. 1853/GB554.
6. Pat1909/GB27449.
7. Gaby Wood, *Living Dolls, A Magical History of the Quest for Mechanical Life*, Faber & Faber, 2002.
8. Now in the Museum of Childhood, Edinburgh.
9. Lydia Richter, *Treasury of Käthe Kruse Dolls*, HP Books, 1982.
10. USD (design)1912/USD43680.
11. Constance Eileen King, *History of Dolls*, Robert Hale, 1977.
12. GB Reg. Design No. 423888, BT51/118.
13. Pauline Cockrill, *The Ultimate Teddy Bear Book*, Dorling Kindersley, 1991.
14. Pat1880/GB1872.
15. Maria Montessori, *Dr Montessori's Own Handbook*. Reprint Schocken, New York, 1965.
16. *Ibid.*
17. Pat1912/GB6706.
18. Pat1914/GB14481.
19. Pat1913/GB17890.
20. Pat1906/GB8118.
21. Pat1913/GB20950.
22. Pat1906/GB6386.
23. Pat1906/GB29186.
24. Pat1901/GB10119.
25. Pat1903/GB7285.
26. Pat 1867/GB1792.
27. Pat 1908/GB10619.
28. Pat1908/GB10618.
29. Pat1903/GB4453.
30. Pat1907/GB2785.

31. Sally Kevill-Davies, *Yesterday's Children*, Antique Collectors Club, 1994.

Chapter Six

1. *Women's Opinion*, January 1874.
2. Pat1874/GB2849.
3. *Woman's Opinion*, 1874, various advertisements.
4. *IRON*, March 1874, *Woman's Opinion*.
5. Ibid., *Woman's Opinion*.
6. Pat1874/GB2849.
7. Pat1878/GB1716.
8. Oldham Local Studies and Archives.
9. *People's Food Leaves*, No. 1.
10. *People's Food Leaves*, No. 8.

Chapter Seven

1. Pat1811/GB3405.
2. Judith Dupré, *Bridges*, Black Dog and Leventhal, New York, 1997.
3. Thomas Cook Archives.
4. Jack Simmons, *The Victorian Railway*, Thames & Hudson, 1991.
5. Pat 1904/GB1567.
6. Pat1913/GB26237.
7. Pat1856/GB791.
8. Pat1907/GB23296.
9. Veronica Davis Perkins, 'Whose Line is it Anyway?' in *Semaphore to Short Waves*, ed. Frank James, RSA, 1998.
10. Public Record Office, 1901 Census England and Wales.
11. Pat1865/GB1628.
12. Pat1859/GB873 and 1859/US23536.
13. Pat1871/US115935.
14. Mario Baraona, *Signal Flares*.
15. Pat1880/GB1601.
16. Pat1909/GB12492.
17. Pat1876/GB3585.
18. Pat1859/GB1014.
19. Pat1863/GB2613.
20. Pat1880/GB2760.
21. Pat1878/GB330.
22. Pat1890/GB12884.
23. Pat1913/GB26136.
24. Pat1907/GB24909.
25. Pat1870/GB2822.
26. Patricia Marks, *Bicycles, Bangs and Bloomers*, University Press of Kentucky, 1990.
27. Pat1886/GB4566.
28. Pat1903/GB15237.
29. Pat1897/GB26054.
30. Pat1897/GB25289.
31. Pat1904/GB10265.
32. Pat1887/GB13125.
33. Pat1897/GB12836.
34. Pat1898/GB15679.
35. Pat1903/US743801.
36. Pat1907/GB23574.
37. Pat1903/GB4925.
38. John Dunlop, a Belfast vet, had invented the pneumatic tyre in 1888.
39. Pat1904/GB21267.
40. Pat1904/GB6519.
41. Pat1905/GB6485.
42. Pat1908/GB12467.
43. Pat1910/GB5908.

Chapter Eight

1. Patricia Phillips, *Scientific Lady*, Weidenfeld & Nicolson, 1990.
2. Pat1872/GB1198.
3. Pat1884/GB5443.
4. Stephen van Dulken, *Inventing the 19th Century*, British Library, 2001.
5. Pat1904/GB9421.
6. Royal Commissioners for the Exhibition of 1851.
7. Pat1904/GB17256.
8. Pat1808/GB3129.
9. Pat1908/GB13084.
10. Pat1875/GB4845.
11. Consumption: TB; phthisis: a disease causing the body to waste like pulmonary TB.
12. Pat1890/GB9622.
13. J.F.I. Caplin, MD, *The Electro-Chemical Bath: Its Use and Effects*.
14. Pat1860/GB656.
15. Hugh Small, *Florence Nightingale: Avenging Angel*, Constable, 1999.
16. Pat1905/GB12428.
17. Pat1907/GB5166.
18. Pat1905/GB6459.
19. Sally Kevill-Davies, *Yesterday's Children*, Antique Collectors Club, 1994.
20. Pat1905/GB12025.

21. Pat1903/GB19157.
22. Pat1908/GB9649.
23. Michael Freeman, *Railways and the Victorian Imagination*, Yale, 1999.
24. Pat1853/GB2368.
25. Pat1862/GB2464.
26. Pat1905/GB11904.
27. Pat1899/GB22790.
28. Pat1903/GB10427.
29. Pat1897/GB256731.
30. Pat1904/GB2639.
31. Pat1908/GB16035.
32. Pat1905/GB10013.
33. Pat1904/GB27650.
34. Pat1904/GB21351.
35. Roy Porter, *The Greatest Benefit to Mankind*, HarperCollins, 1997.
36. Pat1868/GB1602.

Chapter Nine

1. Derek Hudson and Kenneth Luckhurst, *The Royal Society of Arts, K.W. Murray, 1954.*
2. RSA Archives.
3. Fortnum and Mason Archives.
4. Nikolaus Pevsner, *High Victorian Design.*
5. Pat1848/GB12198.
6. *Official Descriptive and Illustrated Catalogue of the Great Exhibition 1851.*
7. Stephen van Dulken, *Inventing the 19th Century.*
8. Asa Briggs, *Victorian Things*, Penguin, 1988.
9. Steven Millhauser, *Martin Dressler, The Tale of an American Dreamer*, Vintage, 1996.
10. Royal Commission for Chicago Exhibition 1893, *Official Catalogue of the British Section*, William Clowes & Sons, 1893.
11. Pat1890/GB5917 and 13903.
12. Pat1889/GB2931.
13. Pat1889/GB4354.

Chapter Ten

1. Pat1859/GB2573.
2. Pat1874/GB607/2849.
3. Pat1878/GB1051.
4. *IRON* as reprinted in *Woman's Opinion*, 28 March 1874.
5. Westminster City Archives: John Burgess and Son Ltd, 1933; *The Grocer*, July 1951; John Burgess and Son Ltd, 1950; Log book of J.H.N. Linklater and Hackwood, 1856–59.
6. The Tussauds Group Chronology.
7. Pam Pilbeam, 'Madame Tussaud and her Waxworks', *History Today*, December 2002.
8. RefAAD 1/29; 1/34; 1/45; –1983 Dora Lunn, Archive of Art and Design, Victoria and Albert Museum.
9. Anthea Callen, *Angel in the Studio.*
10. Judith Walkowitz, *City of Dreadful Delight*, Virago, 1992.
11. Pat1911/GB22232.
12. Pat1908/GB11143.
13. Pat1897/GB26639.
14. Pat1913/GB26989.
15. Pat1870/GB2364 & 871/US116842.
16. Inventors' Museum.
17. Pat1909/GB10531.
18. Pat1904/GB8280.

Chapter Eleven

1. The first London underground train ran on the Metropolitan Railway from Bishop's Road to Farringdon Street in 1863.
2. Judith Walkowitz, *City of Dreadful Delight.*
3. Pat1910/GB1230.
4. Pat1913/GB2873.
5. Pat1904/GB2340.
6. Pat1904/GB25906.
7. Pat1904/GB18590.
8. Pat1897/GB2648.
9. Pat1903/GB6976.
10. Pat1897/GB26416.
11. Pat1903/GB7197.
12. Pat1904/GB18676.
13. Pat1909/GB9944.
14. Pat1876/GB545.
15. Pat1908/GB4016.
16. Pat1907/GB757.
17. Pat1901/GB4453.
18. Nobel Museum.
19. 'The Greenfield Report', *Guardian*, 28 November 2002.

Bibliography

Anscombe, I. *A Woman's Touch: Women in Design from 1860 to the Present Day*. Virago, 1984

Aries, P. *Centuries of Childhood*. Penguin, 1973

Baraona, M. *Signal Flares (website)*.

Barthes, R. *Mythologies*. Granada Publishing Ltd, 1981

Barty-King, H. *Girdle Round the Earth: The Story of Cable and Wireless*. Heinemann, 1979

Beeton, I. *Mrs Beeton's Household Management*. Ward, Lock & Co. 1948

Benjamin W. *Illuminations*. Fontana, 1979

Benson, A and Viscount Esher (eds). *The Letters of Queen Victoria*, vols I, II, III. John Murray, 1908

Bradbury, M. (ed.). *The Atlas of Literature*. De Agostini, 1996

Briggs, A. *Victorian Cities*. Penguin, 1990

—— *Victorian People*. Pelican, 1965

—— *Victorian Things*. Penguin, 1988

—— (ed.). *William Morris: Selected Writings and Designs*. Penguin, 1973

British Library. *Alphabetical Index of Patentees and Applicants for Patents of Invention*. The British Library.

Brontë, C. *Villette*. Penguin, 1979

Bunker, S. *Strawopolis, Luton Transformed 1840–1876*. Bedfordshire Historical Record Society, 1999

Cable and Wireless. *A History of Cable and Wireless*, 1995

Callen, A. *The Angel in the Studio*. Astragal, 1979

Cannadine, D. *Class in Britain*. Penguin, 2000

—— *Ornamentalism*. Penguin, 2002

Caplin, J.F.I, MD. *Selection of Documents and Autograph Letters in Festimony of the Cures effected by the Electrochemical Bath of J.F.I. Caplin*. Translated from the French by H. Baillere, London, 1865

Caplin, R. *Health and Beauty: or, Woman and Her Clothing, considered in relation to The Physiological Laws of the Human Body*. Kent & Co., 1864

Chesney, K. *The Victorian Underworld*. Pelican, 1972

Christies Scotland. *Catalogue*. November 1986

Cockrill, P. *The Ultimate Teddy Bear Book*. Dorling Kindersley, 1991

Colley, L. *Britons: Forging the Nation 1707–1837*. Pimlico, 1994

Commissioners for the Great Exhibition. *Official Descriptive and Illustrated Catalogue of the Great Exhibition 1851*, vols 1, 2, 3. By Authority of the Royal Commission, Spicer Brothers, 1851

Cook, C. *The Longman Companion to Britain in the Nineteenth Century 1815–1914*. Pearson Education, 1999

Davidson, C. *The World of Mary Ellen Best*. Chatto & Windus, 1985

Davies, N. *Europe*. Pimlico, 1997

Davis Perkins, V. Whose line is it anyway? in *Semaphore to Short Waves*, ed. F. James. RSA, 1998

De Vries, L. *History as Hot News 1842–1865*. John Murray, 1995

Dickens, C. *Gone Astray and other Papers from Household Words*. Dent, 1980

Dupré, J. *Bridges*, Black Dog and Leventhal, New York, 1997

Durant, S. *Christopher Dresser*. Academy Editions, 1993

Edwards, C. *Victorian Furntiure, Technology and Design*, Manchester University Press, 1993

Farrell, J. *Umbrellas and Parasols*. Batsford, 1986

Fearnley-Whittingstall, H. (ed.). *101 Un Useless Japanese Inventions*. HarperCollins, 1995

Fletcher, B. *A History of Architecture*. The Athlone Press, 1961

Foreman, A. *Georgiana Duchess of Devonshire*. Flamingo, 1999

Freeman, C. *Luton and the Hat Industry*. The Borough of Luton Musem and Art Gallery, 1996

Freeman, M. *Railways and the Victorian Imagination*. New Haven, CT, Yale University Press, 1999

Gaskell, E. *North and South*. Penguin, 1995

Gathorne-Hardy, J. *The Rise and Fall of the British Nanny*. Weidenfeld & Nicolson, 1985

Girls' Public Day School Trust *1872–1972, A Centenary Review*. GPDST, 1972

Godden, G.A. *Encyclopaedia of British Pottery and Porcelain Marks*. Barrie & Jenkins, 1984

Greer, G. *The Obstacle Race*. Book Club Associates, 1980

Groves, S. *History of Needlwork Tools and Accessories*. Country Life, 1968

Hanaford, P. *1829–1921, Daughters of America*. B.B. Russell, 1983

Hardyment, C. *From Mangle to Microwave*. Polity Press, 1988

—— *Dream Babies: Child Care from Locke to Spock*. Jonathan Cape, 1983

Hart-Davis, A. *What the Victorians Did for Us*. Headline, 2001

Haskell, F. *History and its Images*. New Haven, CT, Yale University Press, 1993

Heskett, J. *Industrial Design*. Thames and Hudson, 1980

Hickman, K. *Daughters of Britannia*. Flamingo, 2000

Hill, C. *Reformation to Industrial Revolution*. Penguin, 1992

Holroyd, M. *Works on Paper*. Little, Brown & Co., 2002

Hosbawm, E.J. *Industry and Empire*. Penguin, 1990

—— and Ranger, T. (eds). *The Invention of Tradition*. Cambridge University Press, 1992

Hudson, D. and Luckhurst, J. *The Royal Society of Arts 1754–1954*. K.W. Murray, 1954

Jaffé, D. *Victoria – A Celebration*. Carlton, 2001

James, F.A.J.L. *Semphores to Short Waves*. RSA, 1998

Jones, D. *Toy with the Idea*. Norfolk Museums Service, 1980

Kelly, A. *Mrs Coade's Stone*. The Self Publishing Group and the Georgian Group, 1990

Kevill-Davies, S. *Yesterday's Children*. Antique Collector's Club, 1994

King, C.E. *The Collector's History of Dolls*. Robert Hale, 1977

Levitt, S. *Victorians Unbuttoned: Registered Design for Clothing, their Makers and Wearers 1839–1900*. HarperCollins, 1986

Levy, M. *Liberty Style*. Weidenfeld & Nicolson, 1986

Lewis, A. *People's Food Leaves*, Nos 1–12, 1878

Lovett Collection of Dolls: Ethnic and Unusual Dolls from Around the World. Museum of Childhood, Edinburgh, Spring 1998

Lurie, A. *The Language of Clothes*. Hamlyn, 1983

McDowell, C. *Status, Style and Glamour*. Thames and Hudson, 1992

Mackenzie, J.M. (ed.). *The Victorian Vision: Inventing New Britain*. V & A Publications, 2001
Marks, L. *Metropolitan Maternity*. Amsterdam, Editions Rodopi B.V., 1996
Marks, P. *Bicycles, Bangs and Bloomers*. University Press of Kentucky, 1990
Mayhew, H. *London Labour and the London Poor*. Penguin, 1985
Millhauser, S. *Martin Dressler, The Tale of an American Dreamer*. Vintage, 1996
Montessori, M. *Dr Montessori's Own Handbook*. Schocken, 1965
Morgan, K. *The Birth of Industrial Britain: Economic Change 1750–1850*. Addison Wesley Longman, 1999
Morris, J. *Pax Britannica*. Faber & Faber, 1998
Mustienes, C. (ed.). *1000 Extraordinary Objects*. Taschen, 2000
Naylor, G. *The Arts and Crafts Movement*. Studio Vista, 1980
Nevill Jackson, F. *Toys of Other Days*. Country Life, 1918
Papanek, V. *Design for the Real World: Making to Measure*. Thames and Hudson, 1972
Parker, R. *The Subversive Stitch*. The Women's Press, 1984
—— and Griselda P. *Old Mistresses: Women, Art and Ideology*. Routledge & Kegan Paul, 1981
Patent Office Illustrated Official Journal (Patents) (1884 onwards). *The Commissioners of Patents' Journal*. Patent specifications published by the British Patent Office.
Perkin, J. 'Sewing Machines: Liberation or Drudgery for Women?', *History Today*, December 2002
Petroski, H. *The Evolution of Useful Things*. Pavilion, 1993
Pevsner, N. *High Victorian Design: a Study of the Exhibits of 1851*. Architectural Press, 1951
—— *A History of Building Types*. Thames & Hudson, 1976
Phillips, P. *The Scientific Lady*. Weidenfeld & Nicolson, 1990
Pilbeam, P. 'Madame Tussaud and her Waxworks', *History Today*, December 2002
Porter, R. *The Greatest Benefit to Mankind: A Medical History of Humanity from Antiquity to the Present*. HarperCollins, 1997
Potter, B. *Peter Rabbit*. Frederick Warne & Co., 2002
Rees, S. *The Floating Brothel*. Headline Review, 2002
Ribeiro, A. *Dress and Morality*. Batsford, 1986
Richter, L. *Treasury of Käthe Kruse Dolls*. HP Books, 1982
Robertson, P. *New Shell Book of Firsts*. Headline, 1994
Robinson, J. *Pandora's Daughters: The Secret History of Enterprising Women*. Constable & Robinson, 2002
Rolt, L.T.C. *Isambard Kingdom Brunel*. Penguin, 1972
Royal Commission. *Great Exhibition of the Works of Industry of All Nations. Official Descriptive and Illustrated Catalogue*, vols I, II, III, 1851
Royal Commission of the Chicago Exhibition, 1893. *Official Catalogue of the British Section*. William Clowes & Sons, 1893
Scott, P. *The Book of Silk*. Thames and Hudson, 1993
Shannon, R. *Gladstone: Peel's Inheritor 1809–1865*. Penguin, 1999
—— *Gladstone: Heroic Minister 1865–1898*. Penguin, 2000
Simkins, M. *An Englishwoman's Home*. Low Marston, 1909
Simmons, J. *The Victorian Railway*. Thames and Hudson, 1991
Small, H. *Florence Nightingale: Avenging Angel*. Constable, 1999
Smith, P. *Disraeli: A Brief Life*. Cambridge University Press, 1999
Strachey, L. *Eminent Victorians*. Penguin, 1986
—— *Queen Victoria*. Harcourt Brace, 1921
Stanley, A. *Mothers and Daughters of Invention: Notes for a Revised History of Technology*. Scarecrow Press, 1993
Summer, L. 'Working Class Women and Corsetry', *Costume*, Number 36 (2002)
Tannahill, R. *Food in History*. Penguin, 1988

Taylor, J., Whalley, J.I., Hobbs, A., Battrick, E.M. *Beatrix Potter 1866–1943: The Artist and her World*. Penguin, 1995
Thompson, E.P. *The Making of the English Working Class*. Pelican, 1979
Tomalin, C. *The Life and Death of Mary Wollstonecraft*. Weidenfeld & Nicolson, 1974
Towler, J. and Bramall, J. *Midwives in History and Society*. Croom Helm, 1986
Trounce, W. *Society of Arts Catalogue of the Seventh Exhibition of Inventions*. 1855
Van Dulken, S. *Inventing the 19th Century; The Great Age of Victorian Inventions*. British Library, 2001
—— *British Patents of Invention 1617–1977*. British Library, 1999
Vare, E.A. and Ptacek, G. *Mothers of Invention; from the bra to the bomb*. Quill, 1989
—— *Patently Female*. John Wiley & Sons, 2002
Vickery, A. *The Gentleman's Daughter: Women's Lives in Georgian England*. Yale University Press, 1998
Walker, H. (ed.). *Cooks and Other People: Proceedings of the Oxford Symposium on Food and Cookery 1995*. Prospect Books, 1996
Walkowitz, J. *City of Dreadful Delight: Narratives of Sexual Danger in Late Victorian London*. Virago, 1992
Wild, A. *The East India Company*. HarperCollins, 1999
Willet Cunnington, C. *English Women's Clothing in the Nineteenth Century*. Dover, 1990
Willet, C. and Cunnington, P. *The History of Underclothes*. Dover, 1992
Wilson, C.A. *Food and Drink in Britain: From the Stone Age to Recent Times*. Penguin, 1984
Wilson, E. *Adorned in Dreams: Fashion and Modernity*. Virago, 1985
Wood, A. *Nineteenth-Century Britain 1815–1914*. Longman, 1982
Wood, G. *Living Dolls, A Magical History of the Quest for Mechanical Life*. Faber & Faber, 2002
Wright, D.G. *Democracy and Reform 1815–1885*. Longman, 1996
Wright, L. *Clean and Decent: The History of the Bathroom and W.C.* Routledge & Kegan Paul, 1984
Wroughton, J. *The Stuart Age 1603–1714*. Longman, 1997
Ziegler, P. *Melbourne*. Collins, 1976

Journals

Costume, various
English Woman's Domestic Magazine, various
'Greenfield Report', *Guardian*, 28 November 2002
The Grocer, July 1950, John Burgess and Son, 1950
History Today, December 2002
Household Journal, various
Illustrated London News, various
Lady's Newspaper, various
Tailor and Cutter, various
The Times, various
Woman's Opinion, various
Trademark Journals, various

Other Sources

Archive of Art and Design, Victoria and Albert Museum
Design Registration Volumes, various. Public Record Office

Inventors' Museum
Nobel Foundation Museum
Public Record Office, 1851–2001 Census England and Wales
Report of Chief Inspector of Factories and Workshops, 1891
The Royal Commissioners for the Exhibition of 1851
The Tussauds Group Chronology
United States Patent and Trademark Office
Westminster City Archives, John Burgess and Son, 1933

The British Patent Office does not possess a public library of its own publications. These, foreign patent publications, are maintained by the British Library at its London site. Their patent information section's web site, http://www.bl.uk/patents, offers links to numerous sites including free databases as well as giving much relevant information. They can be contacted on patents-information@bl.uk or on tel. 020 7412 7919.

To search for British patents back to 1870 and US patents back, by number only, refer to Esp@cenet database at http://gb.espacenet.com.

Index

References to pictures are shown in **bold**.